APPLICATIONS OF HARMONIC MEASURE

The University of Arkansas Lecture Notes in the Mathematical Sciences

Volume 6. Geoffrey S. Watson, *Statistics on Spheres*

Volume 7. Werner C. Rheinboldt, *Numerical Analysis of Parametrized Nonlinear Equations*

Volume 8. John B. Garnett, *Applications of Harmonic Measure*

The first five volumes in the Series were published by the University of Arkansas Department of Mathematics

Volume 1. G. G. Lorentz with a contribution by P. Neval. *Problems in Approximation Theory*

Volume 2. James Glimm and Arthur Jaffe. *Probability Applied to Physics*

Volume 3. Kyoshi Iseki. *Non-Greek Mathematics*

Volume 4. W. A. J. Luxemburg. *Some Aspects of the Theory of Riesz Spaces*

Volume 5. Nathan Jacobson. *Structure Theory of Jordan Algebras*

VOLUME 8

APPLICATIONS OF HARMONIC MEASURE

JOHN B. GARNETT
University of California, Los Angeles

A WILEY-INTERSCIENCE PUBLICATION
JOHN WILEY & SONS
New York Chichester Brisbane Toronto Singapore

Library of Congress Cataloging-in-Publication Data

Garnett, John B.
Applications of harmonic measure.

(The University of Arkansas lecture notes in the mathematical sciences ; v. 8)
"A Wiley-Interscience publication."
Bibliography: p.
1. Measure theory. 2. Harmonic analysis.
I. Title. II. Series.
QA312.G318 1987 515.4′2 87-6123
ISBN 0-471-62772-0

Printed in United States of America

10 9 8 7 6 5 4 3 2 1

Preface

Our purpose here is to illustrate how elementary harmonic measure arguments have broad applications. The introduction gives all the background needed for following these notes, but the reader familiar with bits of [2], [16], [17] or [44] will have an advantage.

Let me thank the University of Arkansas Mathematics Department for its wonderful hospitality and the charming city of Fayetteville for its dogwood blossoms. Thanks also to the audience for finding so many mistakes.

I am most grateful to Laurie Beerman for another perfect typing job.

Part of the work described below was supported by the NSF grant MCS 8002955.

JOHN GARNETT

Los Angeles
March 4, 1986

Contents

APPLICATIONS OF HARMONIC MEASURE

1. INTRODUCTION

Let Ω be a domain in the plane $\mathbb{C}$ or in the sphere $\mathbb{C} \cup \{\infty\}$, and assume every component of the boundary $\partial\Omega$ is a continuum. For a Borel subset $E \subset \partial\Omega$ we define the <u>harmonic</u> <u>measure</u> of E at $z \in \Omega$ (relative to Ω) as the Perron solution to the Dirichlet problem with boundary value the characteristic function χ_E of E:

$$\begin{aligned} \omega(z,E,\Omega) &= \omega(z,E) \qquad (1.1) \\ &= \sup\{v(z) : v \text{ subharmonic in } \Omega,\ \overline{\lim_{z\to\zeta}}\, v(z) \le \chi_E(\zeta), \\ &\qquad \zeta \in \partial\Omega\}. \end{aligned}$$

Ahlfors [1] gives an introduction to the Perron method and the proof that

a) $z \to \omega(z,E)$ is harmonic in Ω.

We note three easy consequences of the definition (1.1):

b) $0 \le \omega(z,E) \le 1$

c) $\omega(z,E)$ increases with E,

$$E_1 \subset E_2 \Rightarrow \omega(z,E_1) \le \omega(z,E_2)$$

d) $\omega(z,E,\Omega)$ increases as Ω increases with $E \subset \partial\Omega$,

$$\Omega_1 \subset \Omega_2,\ E \subset \partial\Omega_1 \cap \partial\Omega_2 \Rightarrow$$

$$\omega(z,E,\Omega_1) \le \omega(z,E,\Omega_2), \quad z \in \Omega_1.$$

We call c) and d) <u>comparisons.</u> Write E^0 for the interior of E relative to $\partial\Omega$ and write $\overline{E}$ for the closure of E. Then

e) $$\lim_{z\to\zeta} \omega(z,E) = \begin{cases} 1, & \zeta \in E^0 \\ 0, & \zeta \notin E \end{cases}$$

This holds because, by our assumptions on Ω, every point of $\partial\Omega$ is a regular point for the Dirichlet problem [44, p. 7].

The set function $E \to \omega(z,E)$, for fixed $z \in \Omega$, is a Borel measure on $\partial\Omega$. If $f(\zeta)$ is continuous on $\partial\Omega$, then the integral

$$u(z) = \int_{\partial\Omega} f(\zeta) d\omega(z,\zeta)$$

is the solution to the Dirichlet problem on Ω for the boundary values $f(\zeta)$. A proof is given in the Appendix below.

Harmonic measure is important in function theory because it is <u>conformally invariant</u>. If $\varphi : \Omega_1 \to \Omega_2$ is analytic, one-to-one, and onto and if φ extends to a continuous map from $\bar{\Omega}_1$ onto $\bar{\Omega}_2$, which necessarily maps $\partial\Omega_1$ onto $\partial\Omega_2$, then for $E \subset \partial\Omega_2$,

$$\omega(z, \varphi^{-1}(E), \Omega_1) = \omega(\varphi(z), E, \Omega_2), \tag{1.2}$$

by the definition (1.1).

<u>Example 1</u>. If Ω is the unit disc $D = \{z : |z| < 1\}$, then

$$\omega(z_0, E) = \frac{1}{2\pi} \int_E \frac{1 - |z_0|^2}{|e^{i\theta} - z_0|^2} \, d\theta.$$

This formula follows from the case $z_0 = 0$ by (1.2) for $\varphi(z) = (z - z_0)/(1 - \bar{z}_0 z)$.

<u>Example 2</u>. If Ω is the upper half plane $\mathcal{U} = \{y > 0\}$, then

$$\omega(x + iy, E) = \frac{1}{\pi} \int_E \frac{y}{(x - t)^2 + y^2} \, dt.$$

If E is the segment $\{a \leq t \leq b\}$, then

$$\omega(x + iy, E) = \frac{1}{\pi}\left\{\arc\tan\left(\frac{x - b}{y}\right) - \arc\tan\left(\frac{x - a}{y}\right)\right\}$$

and the level sets $\omega(z,E) = \alpha$, $0 < \alpha < 1$, are arcs of circles centered on the orthogonal line $x = (a + b)/2$.

Example 3. For $R > 1$ let $\Omega = \Omega_R$ be the annulus $\{z : 1 < |z| < R\}$ and let E be the circle $\{|z| = R\}$. Then

$$\omega(z,E,\Omega) = \log|z|/\log R.$$

If $\tilde{\Omega}$ is a topological annulus bounded by two disjoint analytic Jordan curves γ_1 and γ_2, then $\tilde{\Omega}$ is conformally equivalent to Ω_R for some unique R. To determine the value R, set

$$u(z) = \omega(z,\gamma_2,\tilde{\Omega})$$

and

$$a = \frac{1}{2\pi}\int_{\gamma_1} \frac{\partial u}{\partial n}\, ds$$

where the normal vector n is directed into $\tilde{\Omega}$. Then

$$a = \frac{1}{\log R}$$

because the integral defining a is conformally invariant. This reasoning can be extended to prove $\tilde{\Omega}$ is conformally equivalent to some Ω_R [16].

Example 4. Ω is the slit plane $\mathbb{C}\backslash[0,\infty)$. Then $\varphi(z) = z^2$ is a conformal map from the half plane $\mathcal{U}$ onto Ω and φ is continuous from $\partial\mathcal{U}$ to $\partial\Omega$. Hence

$$\omega(z,E,\Omega) = \frac{1}{\pi}\int_F \frac{v}{(u - t)^2 + v^2}\, dt$$

where $u + iv = z^2$ and $F = \{t \in \mathbb{R} : t^2 \in E \subset \partial\Omega\}$. Note that in this example φ is two-to-one on $\partial\mathcal{U}\setminus\{\infty\}$.

The best intuition of harmonic measure comes from Brownian motion. Consider a random continuous path B_t in the plane, parametrized by time and starting at a point z_0. Suppose that with probability one, B_t will hit any circle $\Gamma_r(z_0) = \{z : |z - z_0| = r\}$ in finite time and that the location of its first hit on $\Gamma_r(z_0)$ is uniformly distributed

$$\text{Prob(first hits } \Gamma_r(z_0) \text{ in } E) = \frac{1}{2\pi}\int_E d\theta.$$

Suppose also that the increments $B_{t+s} - B_s$ are independent of s and of B_s and that $B_{t+s} - B_s$ has the same probability distribution as B_t. That is a rough description of Brownian motion, up to a scale of time. A famous theorem of Kakutani is that

$$\omega(z,E,\Omega) = \text{Prob(Brownian motion starting at } z \in \Omega \text{ first hits } \partial\Omega \text{ in the set } E).$$

Kakutani's theorem makes the comparison principles c) and d) easy to remember. See [36] or [37] for rigorous discussions of Brownian motion; it will not be essential in these lectures.

2. LENGTH SUMS

Let D denote the open unit disc in the complex plane and let $\{\bar{D}_j\}$ be a sequence of pairwise disjoint closed discs contained in D. Our first theorem shows how harmonic measures can control $\Sigma \operatorname{diam}(D_j)$. Theorems 2 and 3 below are applications of this result.

Theorem 1. *Set* $\Omega_j = D \setminus \bigcup_{k \neq j} \bar{D}_k$. *If*

$$\inf_j \inf_{z \in \partial D_j} \omega(z, \partial D, \Omega_j) = a > 0 \tag{2.1}$$

then

$$\Sigma \operatorname{diam}(D_j) \leq c(a) < \infty \tag{2.2}$$

where $c(a)$ *depends only on* a.

In other words, if the D_j are so thinly distributed that, uniformly in j, Brownian motion started anywhere on ∂D_j has a positive chance of reaching ∂D before hitting any other ∂D_k, then (2.2) holds. Two proofs will be given, one without and one with Brownian motion.

For the first proof we need the *pseudohyperbolic metric* on D,

$$\rho(z,w) = \left| \frac{z - w}{1 - \bar{z}w} \right|.$$

It is a metric on D and it is conformally invariant:

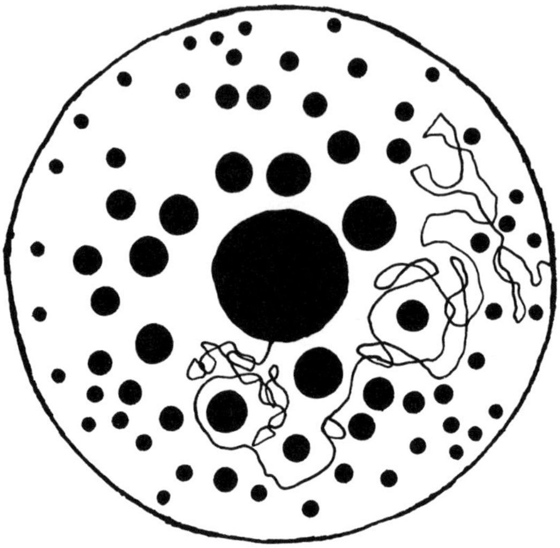

Figure 1: The discs $D_j = \{\rho(z,z_j) < \delta_j\}$ are shaded, and $\delta_j \leq 1/2$. A Brownian path escapes them.

$$\rho(Tz,Tw) = \rho(z,w)$$

if

$$Tz = e^{i\theta}\left(\frac{z-\alpha}{1-\bar{\alpha}z}\right), \quad \alpha \in D.$$

Thus the hyperbolic ball $\Delta = \{z : \rho(z,z_0) < r\}$, $0 < r < 1$, is an euclidean disc having euclidean radius

$$R = \frac{r(1-|z_0|^2)}{1-r^2|z_0|^2}, \tag{2.3}$$

as the reader may verify. Note that if $r \leq 1/2$, then by (2.3)

$$c_1 \operatorname{diam} \Delta \leq r(1-|z_0|) < c_2 \operatorname{diam} \Delta$$

and

$$c_3 \leq \frac{\operatorname{dist}(\Delta, \partial D)}{1-|z_0|} \leq c_4.$$

Conversely, any closed disc $\overline{\Delta} \subset D$ has the form $\{z : \rho(z,z_j) \leq r\}$ for some $r < 1$ because there is $z_0 \in \Delta$ such that $Tz = (z - z_0)/(1 - \overline{z}_0 z)$ maps the annulus $D\backslash\overline{\Delta}$ onto a concentric annulus.

Proof of Theorem 1. We can assume $\{D_j\}$ is a finite set. Replace D_j by a concentric disc with one third the radius. Then (2.1) still holds, but now

$$\operatorname{diam}(D_j) \leq \operatorname{dist}(D_j, \partial D).$$

Hence

$$D_j = \{z : \rho(z,z_j) < \delta_j\}$$

and by (2.3), $\delta_j \leq 1/2$. Also, by the remarks following (2.3) we have

$$c_1' \operatorname{diam}(D_j) \leq \delta_j \operatorname{dist}(D_j, \partial D) \leq c_2' \operatorname{diam}(D_j), \tag{2.4}$$

for constants c_1' and c_2'.

Fix $A > 1$, to be determined later, set

$$\overline{\Delta}_j = \{z : \rho(z,z_j) \leq \delta_j^A\} \subset D_j$$

and consider

$$u(z) = \omega(z, \partial D, D\backslash \cup \overline{\Delta}_j).$$

Because the D_j are disjoint we may, removing one D_j if necessary, assume $0 \notin \cup \overline{D}_j$. Since $\{D_j\}$ is finite, Green's theorem applies to $D\backslash \cup \overline{\Delta}_j$ giving

$$u(0) = \frac{1}{2\pi}\int_{\partial D} u(e^{i\theta})d\theta - \sum_j \frac{1}{2\pi}\int_{\partial\Delta_j} \frac{\partial u}{\partial n} \log\frac{1}{|z|}\, ds,$$

where the normal derivatives are exterior to Δ_j. Hence

$$\sum_j \left\{ \inf_{\partial\Delta_j} \left(\log \frac{1}{|z|} \right) \right\} \left\{ \frac{1}{2\pi} \int_{\partial\Delta_j} \frac{\partial u}{\partial n} \, ds \right\} \leq \frac{1}{2\pi} \int_{\partial D} u(e^{i\theta}) d\theta - u(0) \leq 1.$$

But by (2.4)

$$\inf_{\partial\Delta_j} \log \frac{1}{|z|} \geq \operatorname{dist}(D_j, \partial D) \geq c_1 \delta_j^{-1} \operatorname{diam}(D_j),$$

and so the theorem will be proved if we can show

$$\frac{1}{2\pi} \int_{\partial\Delta_j} \frac{\partial u}{\partial n} \, ds \geq c_1(a) \delta_j. \tag{2.5}$$

To prove (2.5) we use the functions

$$u_j(z) = \omega(z, \partial D, D \setminus \bigcup_{k \neq j} \bar{\Delta}_k)$$

and

$$v_j(z) = \omega(z, \partial\Delta_j, D \setminus \bar{\Delta}_j).$$

By the maximum principle,

$$u(z) \geq u_j(z) - v_j(z), \qquad z \in D \setminus \bigcup \bar{\Delta}_k$$

Let $z \in \partial D_j$. Then by the hypothesis (2.1) and a comparison,

$$u_j(z) \geq a,$$

while

$$v_j(z) = \frac{\log 1/\rho(z, z_j)}{A \log 1/\delta_j} = \frac{1}{A}.$$

Taking $A > 2/a$, we obtain

$$u(z) \geq a/2, \qquad z \in \partial D_j. \tag{2.6}$$

Let V_j be the annulus $D_j \setminus \bar{\Delta}_j$. Then by (2.6),

$$u(z) \geq \frac{a}{2}\, \omega(z, \partial D_j, V_j)$$

on V_j. Hence

$$\frac{\partial u}{\partial n} \geq \frac{\partial \omega(z, \partial D_j, V_j)}{\partial n}, \qquad z \in \partial\Delta_j.$$

By Example 3 above

$$\frac{1}{2\pi}\int_{\partial\Delta_j} \frac{\partial \omega(z, \partial D_j, V_j)}{\partial n}\, ds = \frac{1}{(A-1)\log 1/\delta_j},$$

and since $\delta_j \leq 1/2$, that gives

$$\frac{1}{2\pi}\int_{\partial\Delta_j} \frac{\partial u}{\partial n}\, ds \geq \frac{1}{(A-1)\log 1/\delta_j} \geq c_1(a)\delta_j,$$

which is (2.5). □

We now sketch the Brownian motion proof; it is due to Burgess Davis. Let P_z denote probability on the space of Brownian paths B_t starting at z and let E stand for integration with respect to P_0. By Kakutani's theorem the measurable set

$$A_j = \{B_t \text{ hits } \overline{D}_j \text{ before hitting } \partial D\}$$

satisfies

$$P_0(A_j) \geq \frac{\operatorname{diam}(D_j)}{2\pi} \tag{2.7}$$

Let $N = \Sigma\, \chi_{A_j}$. The random variable (or measurable function) N counts the number of $\overline{D}_j$ visited by B_t before B_t reaches ∂D. By (2.7) we have

$$\Sigma \operatorname{diam}(D_j) \leq 2\pi E(N).$$

The hypothesis of Theorem 1 is that

$$\inf_{z\in\partial D_j} P_z(B_t \text{ hits } \bigcup_{k\neq j} \overline{D}_k \text{ before hitting } \partial D) \tag{2.8}$$
$$= \inf_{z\in\partial D_j} P_z(\bigcup_{k\neq j} A_k) \leq 1 - a < 1.$$

Now (2.8) easily gives a bound for $E(N)$. Let $\tau_1 = \inf\{t : B_t \in \cup \overline{D}_j\}$ be the first time B_t hits $\cup \overline{D}_j$. Assuming that τ_n has been defined and that $B_{\tau_n} \in \overline{D}_{j(n)}$, let

$$\tau_{n+1} = \inf\{t > \tau_n : B_t \in \bigcup_{j\neq j(n)} \overline{D}_j\}$$

be the time of the next visit to a different D_j. The functions τ_n and $\tau_D = \inf\{t : B_t \in \partial D\}$ are measurable with respect to P_0. Set

$$M = \max\{n : \tau_n \leq \tau_D\}$$

Then $M - 1$ counts the number of trips from one D_j to another taken by B_t before it departs D. Hence $M \geq N$ and

$$E(N) \leq E(M) = \sum_{n=1}^{\infty} P_0(M \geq n).$$

Now by (2.8) and the independence of Brownian increments,

$$\begin{aligned} P_0(M \geq n + 1) &= \sum_j P_0(M \geq n + 1, B_{\tau_n} \in \overline{D}_j) \\ &\leq (1 - a) \sum_j P_0(M \geq n, B_{\tau_n} \in \overline{D}_j) \\ &= (1 - a)P_0(M \geq n), \end{aligned}$$

so that by induction

$$\sum \operatorname{diam} D_j \leq 2\pi \sum_{n=1}^{\infty} (1 - a)^{n-1} = 2\pi/a.$$

Both proofs of Theorem 1 have advantages. The Brownian motion proof is more conceptual and, even when the probabilistic statements are made rigorous, less technical. The first proof is easier to generalize to certain settings, such as the Laplace-Beltrami equation on the complex unit ball.

Theorem 1 is from [18]. It is related to Hall's lemma [22], [16] and to earlier work by Beurling [6] and Nevanlinna [33]. See [11] and [17] for other variants of Hall's lemma. The converse of Theorem 1 is false; Beurling [7] gives a counterexample.

3. LEVEL CURVES OF CONFORMAL MAPPINGS

As an application of Theorem 1 we prove

Theorem 2 (Hayman, Wu). *Let* Ω *be a simply connected plane domain, let*

$$\varphi : D \to \Omega$$

be a conformal mapping from the unit disc D *onto* Ω, *and let* L *be any line in the plane. Then* $\varphi^{-1}(\Omega \cap L)$ *is a countable union of rectifiable arcs* Γ_j *and*

$$\sum \text{length}(\Gamma_j) \le C_0$$

where C_0 *is a constant independent of* Ω, *of* φ, *and of* L.

For the proof we need two elementary geometric properties of the conformal mapping φ,

$$\frac{|\varphi'(z)|(1 - |z|^2)}{4} \le \text{dist}(\varphi(z), \partial\Omega) \le |\varphi'(z)|(1 - |z|^2) \tag{3.1}$$

and, if $\rho(z_1, z_2) = \delta$,

$$\frac{\delta}{(1 + \delta)^2} \le \frac{|\varphi(z_1) - \varphi(z_2)|}{\text{dist}(\varphi(z_1), \partial\Omega)} \le \frac{4\delta}{(1 - \delta)^2}. \tag{3.2}$$

Let $w_0 = \varphi(z_0)$ and $d = \text{dist}(w_0, \partial\Omega)$. Then the right-hand inequality of (3.1) is Schwarz's lemma at $w = 0$, applied to the function

$$g(w) = \frac{\varphi^{-1}(w_0 + dw) - z_0}{1 - \bar{z}_0\varphi^{-1}(w_0 + dw)}$$

For the left side of (3.1) note that

$$h(z) = \frac{\varphi\left(\dfrac{z + z_0}{1 + \bar{z}_0 z}\right) - w_0}{\varphi'(z_0)(1 - |z_0|^2)}$$

is a normalized univalent function, $h(0) = 0$, $h'(0) = 1$, and apply the Koebe theorem that $h(D) \supset \{w : |w| < 1/4\}$. And (3.2) follows from (3.1) and the distortion theorem

$$\frac{|z|}{(1 + |z|)^2} \le |h(z)| \le \frac{|z|}{(1 - |z|)^2}$$

for normalized univalent functions.

Partition $L \cap \Omega$ into segments L_j so that

$$c_1 \le \frac{\text{length}(L_j)}{\text{dist}(L_j, \partial\Omega)} \le 1/2.$$

Let $\Gamma_j = \varphi^{-1}(L_j)$, let w_j be the midpoint of L_j and let $z_j = \varphi^{-1}(w_j)$. Then by (3.2) there are δ_1 and δ_2, $0 < \delta_2 < \delta_1 < 1$, such that

$$\Gamma_j \subset \{z : \rho(z, z_j) \le \delta_1\} \tag{3.3}$$

and

$$\rho(z_j, z_k) \ge \delta_2, \quad k \ne j. \tag{3.4}$$

Also, by (3.1)

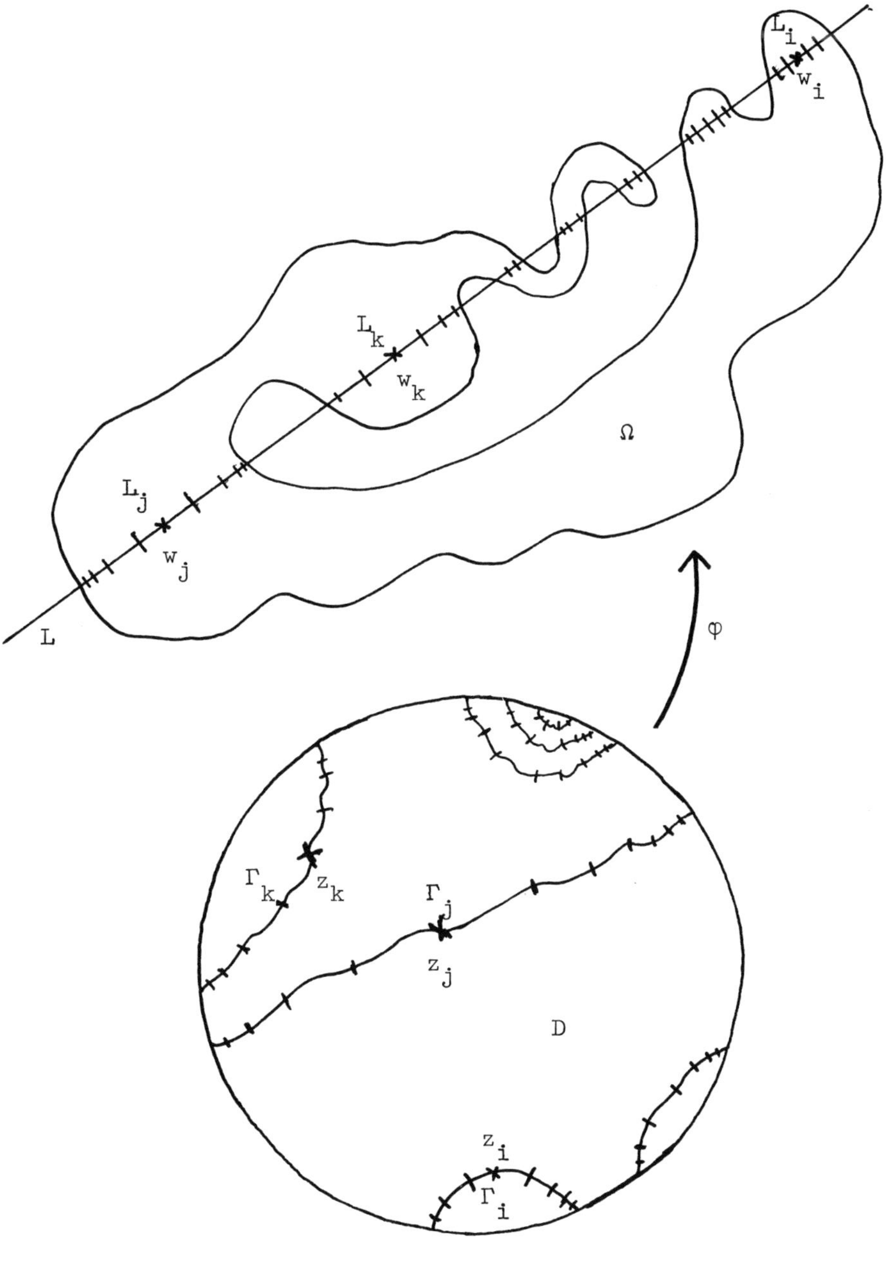

Figure 2: The L_j and Γ_j.

$$\begin{aligned} \text{length}(\Gamma_j) &= \int_{L_j} \left|\frac{dz}{dw}\right| |dw| \\ &\le \int_{L_j} \frac{1 - |\varphi^{-1}(w)|^2}{\text{dist}(w, \partial\Omega)} |dw| \\ &\le \frac{c_2(1 - |z_j|^2)\text{length}(L_j)}{\text{dist}(L_j, \partial\Omega)} \le c_3(1 - |z_j|^2), \end{aligned} \tag{3.5}$$

because by (3.4) $\sup_{\Gamma_j} (1 - |z|^2) \le c(1 - |z_j|^2)$.

Now let $\delta < \delta_2/4$ be a small number to be determined later and set

$$D_j = \{z : \rho(z, z_j) < \delta\}.$$

The discs D_j are pairwise disjoint by (3.4) and $\text{diam}(D_j) \le c(\delta)(1 - |z_j|)$. So by (3.5) and Theorem 1, the theorem will be proved if $u_j(z) = \omega(z, \partial D, D \setminus \bigcup_{k \ne j} \overline{D}_k)$, satisfies

$$\inf_j \inf_{z \in \partial D_j} u_j(z) \ge a > 0. \tag{3.6}$$

Since $u_j(z)$ is harmonic on $\{z : \rho(z, z_j) < 2\delta\}$, (3.6) will follow from

$$u_j(z_j) \ge a' > 0. \tag{3.7}$$

To prove (3.7) we return to Ω. Assume $L = \mathbb{R}$ and assume $w_j = 0$ for some fixed j. Set

$$K_j = L \cap \Omega \cap \{w : |w| \ge \alpha \, \text{dist}(w_j, \partial\Omega)\}$$

with α so small that $L_k \in K_j$, $k \ne j$, and let

$$v_j(w) = \omega(w, \partial\Omega, \Omega \setminus K_j).$$

Comparing v_j to a harmonic measure on a half disc, we have

$$v_j(w) \leq c \frac{|w - w_k|}{\text{length}(L_k)}$$

if $|w - w_k| < \text{length}(L_k)/2$, so that by (3.2), $v_j(\varphi(z)) \leq c_4\delta$ on $\bigcup_{k \neq j} \overline{D}_k$. Consequently

$$u_j(z) \geq \frac{v_j(\varphi(z)) - c_4\delta}{1 - c_4\delta} \tag{3.8}$$

on $(D \setminus \bigcup_{k \neq j} D_k) \cap \varphi^{-1}(\Omega \setminus K_j)$.

Let $\tilde{\Omega}$ be that component of $\Omega \cap \{\overline{w} : w \in \Omega\}$ such that $0 = w_j \in \tilde{\Omega}$, and let

$$U_j(w) = \omega(w, \partial\tilde{\Omega}, \tilde{\Omega} \setminus K_j).$$

Then by symmetry $v_j(w) + v_j(\overline{w}) \geq U_j(w)$, and thus

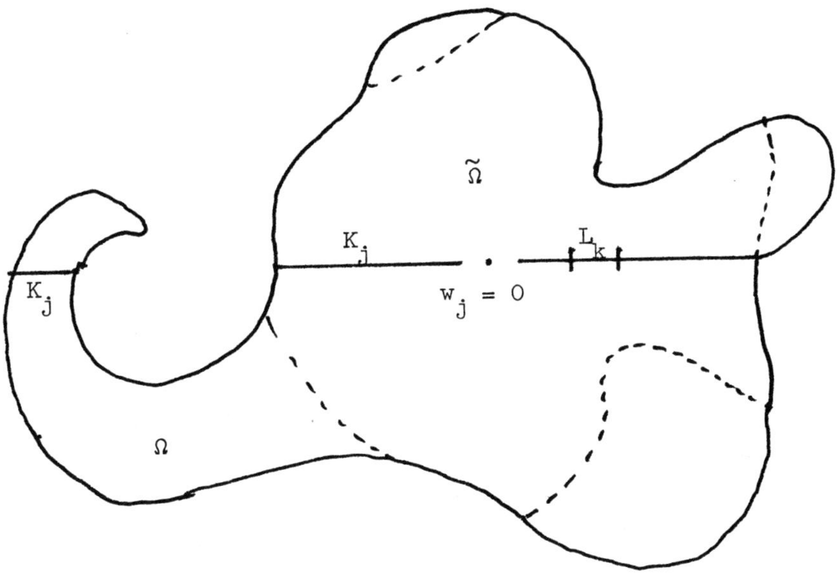

Figure 3: The domains Ω and $\tilde{\Omega}$.

$$v_j(0) \geq U_j(0)/2. \tag{3.9}$$

If $w = \psi(z)$ is the conformal mapping of D onto $\tilde{\Omega}$ with $\psi(0) = 0$ and $\psi'(0) > 0$, then $\psi(\bar{z}) = \overline{\psi(z)}$ and

$$\tilde{\Omega} \cap K_j \subset \psi([-1,1)),$$

and by (3.2) there is $\beta = \beta(\alpha)$ such that

$$K_j \subset \psi([-1,-\beta] \cup [\beta,1]).$$

Hence by conformal invariance,

$$U_j(0) = U_j(\psi(0)) \geq \omega(0, \partial D, D\backslash\{\beta < |x| < 1\})$$
$$\geq c_5\beta,$$

and with (3.8) and (3.9), this gives

$$u_j(z_j) \geq U_j(0)/2 - c_4\delta$$
$$\geq \frac{c_5\beta}{2} - c_4\delta \geq a' > 0$$

if δ is sufficiently small. □

Theorem 2 is due to Hayman and Wu [23]; the proof given above is from [18].

Here are two open problems suggested by Theorem 2. Let Ω be simply connected and let $\psi = \varphi^{-1}$ be a conformal mapping from Ω onto D. Then by Theorem 2,

$$\int_{L\cap\Omega} |\psi'|\,ds \leq c_0$$

for any line L. Baernstein [5] has asked if, for $1 < p < 2$,

$$\int_{L\cap\Omega} |\psi'|^p\,ds \leq c_p.$$

Counterexamples exist for $p = 2$. The second problem comes from Brennan's work [10]. For $4/3 < p < 4$ is it true that

$$\iint_{\Omega} |\psi'|^p \, dxdy < \infty ?$$

Brennan has proved this for $4/3 < p < 3 + \varepsilon$.

4. INTERPOLATING SEQUENCES

The hypothesis (2.1) of Theorem 1 is invariant under conformal self maps of D, but the stated conclusion (2.2) is not. Thus the theorem also has an invariant conclusion. One consequence of the invariant conclusion is a theorem about H^∞ interpolation.

Let $\delta > 0$ be small and suppose $\{z_j\}$ is a sequence in D such that the hyperbolic discs

$$D_j = \{z : \rho(z,z_j) \leq \delta\}$$

are pairwise disjoint. Then

$$\operatorname{diam}(D_j) \geq c_1(\delta)(1 - |z_j|).$$

For fixed z_k, the rotated discs

$$D_{j,k} = \left\{z : \frac{z - z_k}{1 - \bar{z}_k z} \in D_j\right\}$$

have hyperbolic centers $(z_j - z_k)/(1 - \bar{z}_k z_j)$ and the same hyperbolic radius δ and

$$\operatorname{diam}(D_{j,k}) \geq c_1(\delta)\left(1 - \left|\frac{z_j - z_k}{1 - \bar{z}_k z_j}\right|\right).$$

So if

$$\inf_j \omega(z_j, \partial D, D\setminus \bigcup_{k\neq j} D_k) = a > 0,$$

then for all k,

$$\sum_j \left\{1 - \left|\frac{z_j - z_k}{1 - \bar{z}_k z_j}\right|\right\} \leq c_2(a),$$

or

$$\inf_k \prod_{j, j \neq k} \left| \frac{z_j - z_k}{1 - \bar{z}_k z_j} \right| \geq \eta > 0. \tag{4.1}$$

Now (4.1) is Carleson's necessary and sufficient condition that every interpolation problem

$$f(z_j) = \alpha_j, \quad j = 1,2,\ldots$$

with $\{\alpha_j\} \in \ell^\infty$, has solution $f(z)$ bounded and analytic in D; i.e. that $\{z_j\}$ is an interpolating sequence for $H^\infty(D)$.

Theorem 3. _Let_ $\{z_j\}$ _be a sequence in_ D. _Assume that for each_ j _there exists_ $u_j(z)$ _harmonic in_ D _such that_

$$u_j(z_j) = 1$$
$$u_j(z_k) < 0, \quad k \neq j$$

and

$$|u_j(z)| \leq M$$

where M _is independent of_ j. _Then_ $\{z_j\}$ _is an interpolating sequence for_ $H^\infty(D)$.

Indeed, by these hypotheses and Harnack's inequality, there is $\delta = \delta(M) > 0$ such that if $D_j = \{z : \rho(z,z_j) \leq \delta\}$, then

$$u_j(z) > 2/3, \quad z \in D_j$$
$$u_j(z) < 1/3, \quad z \in \bigcup_{k \neq j} D_k.$$

Then the $\{D_j\}$ are disjoint and

$$\omega(z, D \setminus \bigcup_{k \neq j} D_k, \partial D) \geq \frac{u_j(z) - 1/3}{M - 1/3}$$

so that

$$\inf_{z \in \partial D_j} \omega(z, D \setminus \bigcup_{k \neq j} D_k, \partial D) \geq \frac{2}{3M - 1} = a.$$

See [17], Chapters VII and X, and their references for some variations on Theorem 3, and see [18] for further invariant consequences of Theorem 1, including the fact that from Theorem 2, arc length on $\varphi^{-1}(\Omega \cap L)$ is a Carleson measure.

A difficult open problem is to show that a sequence $\{z_j\}$ in the unit ball of $\mathbb{R}^N$, $N > 2$, is an interpolating sequence for the bounded harmonic functions on the ball if $\{z_j\}$ satisfies the analogue of the hypotheses of Theorem 3. See [14].

5. NONTANGENTIAL LIMIT SETS

In $\mathbb{R}^N$, $N \geq 2$, we consider transformations of four types

(i) translations $x \to x + a$, $a \in \mathbb{R}^N$

(ii) dilations $x \to \lambda x$, $\lambda > 0$

(iii) proper rotations $x \to Ax$, A orthogonal and $\det A = 1$

(iv) inversion $x \to Jx = x/|x|^2$, $x \neq 0$.

A (sense preserving) Möbius transformation is a product $T = A_1 A_2 \cdot \cdots \cdot A_n$ where each A_j is of one of the four types above and where $A_j = J$ for an <u>even</u> number of j. Möbius transformations preserve angles and they map spheres to spheres or to hyperplanes (spheres through ∞). The familiar Möbius transformations in $\mathbb{C} = \mathbb{R}^2$ are generated in the same manner and thus each Möbius transformation of the (x_1, x_2) plane extends to a Möbius transformation of $\mathbb{R}^N$ fixing the (x_1, x_2) plane.

Let $B = \{x \in \mathbb{R}^N : |x|^2 = \Sigma x_j^2 < 1\}$ be the unit ball of $\mathbb{R}^N$ and let

$$\mathfrak{M} = \{T : T(B) = B\}$$

be the group of Möbius transformations which keep B invariant. Each $T \in \mathfrak{M}$ also maps $\partial B = S^{N-1}$ bijectively onto itself.

The hyperbolic metric on B is defined by

$$ds = \frac{2|dx|}{1 - |x|^2},$$

so that the distance from $x \in B$ to $y \in B$ is the infimum, over all smooth arcs γ in B joining x to y, of

$$\int_\gamma \frac{2|dz|}{1 - |z|^2}.$$

For example,

$$d(x,0) = \log\left(\frac{1 + |x|}{1 - |x|}\right)$$

since $[0,x]$ is the minimizing arc. The hyperbolic metric is invariant under $\mathfrak{M}$

$$d(Tx,Ty) = d(x,y), \qquad T \in \mathfrak{M}. \tag{5.1}$$

When $N = 2$, (5.1) follows from the identity

$$|T'(z)| = \frac{1 - |T(z)|^2}{1 - |z|^2}$$

and the same identity holds for all N [4]. For $N = 2$ the hyperbolic metric and the pseudohyperbolic metric are related by

$$d = \log\left(\frac{1 + \rho}{1 - \rho}\right). \tag{5.2}$$

Lemma 5.1. *For* $R > 1$ *there is a constant* $c_1(R) < 1$ *such that if* $d(x,y) < R$ *then*

$$|x - y| < c_1(R)(1 - |x|).$$

By considering the plane containing $\{0,x,y\}$, we see it is enough to prove the lemma in the case $N = 2$, where it follows from (2.3) and (5.2).

Write σ for the surface measure on $S^{N-1} = \partial B$, normalized by $\sigma(\partial B) = 1$. For $x \in B$, $|x| > 1/2$ and for $M > 1$, we define the *spherical cap* as

$$C(x,M) = \{z \in \partial B : |z - x| < M(1 - |x|)\}$$

Then there are constants $c_2(M)$ and $c_3(M)$ so that

$$c_2(M)(1 - |x|)^{N-1} \le \sigma(C(x,M)) \le c_3(M)(1 - |x|)^{N-1}. \tag{5.3}$$

For $z \in \partial B$ and $M > 1$ we also define the nontangential *cone* at z to be

$$\Gamma(z,M) = \{x \in B : |x| > 1/2, \ |x - z| < M(1 - |x|)\}$$

Then

$$x \in \Gamma(z,M) \Leftrightarrow z \in C(x,M) \tag{5.4}$$

A subgroup G of $\mathfrak{M}$ is <u>discontinuous</u> if each $x \in B$ has a neighborhood V_x such that

$$T(V_x) \cap V_x = \emptyset, \ T \in G\backslash\{\text{identity}\}.$$

If G is discontinuous then G is countable, $G = \{T_n\}$ and every limit point of $\{T_n x\}$, $x \in B$, is on ∂B. We want to study the set of nontangential limit points for a discontinuous group G. Fix $x_0 \in B$, fix $M > 1$, and set

$$\Lambda(x_0,M) = \bigcap_{k=1}^{\infty} \bigcup_{n=k}^{\infty} C(T_n x_0, M).$$

Thus $z \in \Lambda(x_0,M)$ if and only if z lies in infinitely many caps $C(T_n x_0,M)$, and by (5.4) this happens if and only if $\{T_n x_0\}$ has a subsequence contained in $\Gamma(z,M)$ and converging to z. If $M' > M$ then $\Lambda(x_0,M') > \Lambda(x_0,M)$, but we have

<u>Lemma 5.2</u>. <u>If</u> $M' > M$ <u>then</u>

$$\sigma(\Lambda(x_0,M')\backslash\Lambda(x_0,M)) = 0.$$

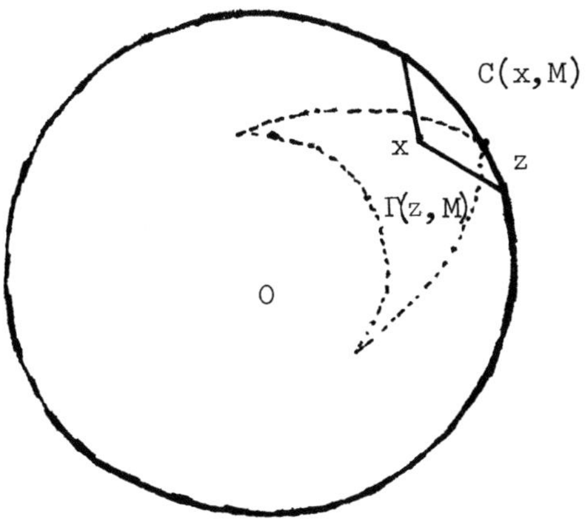

Figure 4: Caps and cones when $M = 2$.

Proof. Set $x_n = T_n x_0$ and

$$E_k = \cap \{\partial B \backslash C(x_n, M) : 1 - |x_n| < 1/k\}.$$

Then $\partial B \backslash \Lambda(x_0, M) = \cup E_k$. Fix k and suppose $z_0 \in \partial B$ is a point of density of E_k,

$$\frac{\sigma(E_k \cap \{|z - z_0| < \delta\})}{\sigma(\{|z - z_0| < \delta\})} \to 1 \qquad (\delta \to 0). \qquad (5.5)$$

We shall show

$$z_0 \notin \Lambda(x_0, M'),$$

which will prove the lemma. If $z_0 \in \Lambda(x_0, M')$, then for any small $\eta > 0$ there is x_n such that

$$1 - |x_n| < 1/k,$$

$$\delta = (M' + M)(1 - |x_n|) < \eta$$

and

$$|x_n - z_0| < M'(1 - |x_n|)$$

Then

$$C(x_n, M) \subset \{|z - z_0| < \delta\} \backslash E_k$$

and

$$\sigma(C(x_n, M)) \geq c_2(M) \left(\frac{\delta}{M' + M} \right)^{N-1}$$

$$\geq c_4(M) \sigma(\{|z - z_0| < \delta\},$$

so that

$$\frac{\sigma(E_k \cap \{|z - z_0| < \delta\})}{\sigma(\{|z - z_0| < \delta\})} < 1 - c_4(M),$$

contradicting (5.5). □

Moreover, if $d(x_0,x_0') = R$, then $d(T_nx_0,T_nx_0') = R$ and by Lemma 5.1,

$$1 - |T_nx_0| < (1 + c_1(R))(1 - |T_nx_0'|) \tag{5.6}$$

Thus there is $M' = M'(M,R)$ such that

$$C(T_nx_0,M) \subset C(T_nx_0',M')$$

and

$$\Lambda(x_0,M) \subset \Lambda(x_0',M').$$

Consequently, the <u>nontangential limit set</u>

$$\Lambda = \Lambda(G) = \bigcup_{M>1} \Lambda(x_0,M)$$

is independent of x_0. And by Lemma 5.2,

$$\sigma(\Lambda) = \sigma(\Lambda(x_0,M))$$

for any x_0 and M. In fact, Λ has the same measure as $\bigcap \{\Lambda(x,M) : x \in B,\ M > 1\}$, so that the largest and smallest choices of the "nontangential limit set" differ only by a set of measure zero. From the definition it is clear that

$$T_n(\Lambda) = \Lambda, \qquad T_n \in G.$$

Fix $G = \{T_n\}$, a discontinuous subgroup of $\mathfrak{m}$. G is <u>of convergence type</u> if

$$\Sigma\,(1 - |T_nx_0|)^{N-1} < \infty$$

By (5.6) this condition is independent of $x_0 \in B$. G is <u>of divergence type</u> if $\Sigma\,(1 - |T_nx_0|)^{N-1} = \infty$.

<u>Theorem 4.</u> <u>If</u> G <u>is of convergence type then</u> $\sigma(\Lambda) = 0$.

<u>Theorem 5.</u> <u>If</u> G <u>is of divergence type then</u> $\sigma(\Lambda) = 1$.

Theorem 4 is very easy. By (5.3) and Lemma 5.2,

$$\sigma(\Lambda) \le \lim_{k\to\infty} \sum_{n=k}^{\infty} \sigma(C(T_n x_0, M))$$
$$\le \lim_{k\to\infty} \sum_{n=k}^{\infty} c_3(M)(1 - |T_n x_0|)^{N-1}.$$

To prove Theorem 5 we need the invariant form of harmonic measure. When $x \in B$ and $z \in \partial B$, we define the <u>Poisson kernel</u>

$$p(x,z) = \left(\frac{1 - |x|^2}{|x - z|^2}\right)^{N-1}$$

and the <u>hyperharmonic measure</u>

$$\omega(x,E) = \int_E p(x,z)d\sigma(z), \qquad E \subset \partial B.$$

Then a change of variable [4] shows

$$\omega(Tx,TE) = \omega(x,E), \qquad T \in \mathfrak{M}.$$

The functions

$$u(x) = \int_{\partial B} f(z)d\omega(x,z)$$

are called hyperharmonic functions. They are annihilated by the invariant Laplacian,

$$Lu = \frac{(1 - |x|^2)^2}{4} \sum_{j=1}^{N} \left(\frac{\partial^2 u}{\partial x_j^2} + \frac{2(N - 2)}{1 - |x|^2} x_j \frac{\partial u}{\partial x_j}\right),$$

which we will not need. If $N \ge 3$, the usual Laplacian is not invariant under $\mathfrak{M}$.

<u>Lemma 5.3</u>. <u>Uniformly in</u> $|x| \ge 1/2$,

$$\lim_{M\to\infty} \omega(x,C(x,M)) = 1.$$

<u>Proof</u>. The angle between the normal to S^{N-1} and the normal to $\{|z - x| = M(1 - |x|)\}$, where these spheres intersect, tends to 0 as $M \to \infty$, uniformly in $|x| \ge 1/2$. That angle is preserved under $T_x \in \mathfrak{M}$ with $T_x(x) = 0$. Consequently the image ball

$B_x = T_x(\{z : |z - x| < M(1 - |x|)\})$ satisfies

$$\inf_{|x|\geq 1/2} \sigma(B_x \cap S^{N-1}) \to 1 \qquad (M \to \infty).$$

But by the invariance of ω,

$$\omega(x, C(x,M)) = \omega(0, B_x) = \sigma(B_x). \qquad \square$$

Lemma 5.4. Either $\sigma(\Lambda) = 0$ or $\sigma(\Lambda) = 1$.

Proof. Assume $\sigma(\Lambda) > 0$ and let $\varepsilon > 0$. Choose M so that by Lemma 5.3,

$$\omega(x, C(x,M)) > 1 - \varepsilon$$

if $|x| > 1/2$. Let z_0 be a point of density of Λ. Then there are $x_k = T_k(0)$ such that $C_k = C(x_k, M))$ has $z_0 \in C_k$ and

$$\frac{\sigma(C_k \cap \Lambda)}{\sigma(C_k)} \to 1 \qquad (k \to \infty). \tag{5.7}$$

But with M fixed we have

$$\sup_{z \in C_k} p(x_k, z) \leq C(M) \inf_{z \in C_k} p(x_k, z)$$

and so (5.7) yields

$$\frac{\omega(x_k, C_k \cap \Lambda)}{\omega(x_k, C_k)} \to 1 \qquad (k \to \infty).$$

Hence there is $T_k \in G$ such that

$$\omega(T_k(0), \Lambda) > 1 - \varepsilon.$$

But then because $T_k(\Lambda) = \Lambda$,

$$\sigma(\Lambda) = \omega(0, \Lambda) = \omega(T_k(0), \Lambda) > 1 - \varepsilon. \qquad \square$$

Proof of Theorem 5. Fix $M > 1$. There is $\delta < 1$, with $1 - \delta$ small, such that for some $a < b$,

$$0 < a < \omega(x,C(x,M)) < b < 1 \tag{5.8}$$

uniformly in $|x| \geq \delta$.

Take $x_0 = 0$ and $x_n = T_n(0)$, where $\{T_n\} = G\backslash\{\text{identity}\}$. After a rotation, T_n is a dilation in the half space model of B ([4] p. 21). Hence there is $M' > 0$ such that if $x_m \neq x_n$, $x_m \neq 0$ and $C(x_m,M) \cap C(x_n,M) \neq \emptyset$ then

$$T_n^{-1}(C(x_m,M)) \subset C(T_n^{-1}x_m,M'). \tag{5.9}$$

We suppose $\sigma(\Lambda) < 1$ and show $\Sigma(1 - |x_n|)^{N-1} < \infty$. By Lemma 5.4, $\sigma(\Lambda) = 0$, and for a fixed ε, $0 < \varepsilon < a$, there exists r, $\delta < r < 1$, such that

$$\sigma\left(\bigcup_{|x_n|>r} C(x_n,M')\right) < \varepsilon < a.$$

Set $R = \log((1 + r)/(1 - r))$ and choose a maximal subsequence $\{y_k\}$ of $\{x_n\}$ such that $y_0 = 0$ and

$$d(y_k,y_j) \geq R, \quad j \neq k.$$

For $y_k \neq 0$ define $C_k = C(y_k,M)$ and $E_k = C_k\backslash D_k$ where

$$D_k = \cup\ \{C_j : |y_j| \geq |y_k|,\ y_j \neq y_k\}.$$

Then $E_k \cap E_j = \emptyset$, $j \neq k$. We claim

$$\sigma(E_k) \geq c(1 - |y_k|)^{N-1}. \tag{5.10}$$

By (5.8) we have $\omega(y_k,C_k) > a > \varepsilon$, while by (5.9)

$$\begin{aligned}\omega(y_k,D_k) &= \omega(0,T_k^{-1}D_k)\\ &\leq \omega\left(0, \bigcup_{j\neq k} C(T_k^{-1}y_j,M')\right) < \varepsilon.\end{aligned}$$

Hence $\omega(y_k,E_k) > a' > 0$, and because $p(y_k,z) \geq c(M)(1 - |y_k|)^{1-N}$ when $z \in C_k$, (5.10) holds.

Now because G is discontinuous, each hyperbolic ball $\{x : d(x,y_k) < R\}$ contains at most $K = K(R)$ orbit points x_n, and because $\{y_k\}$ is maximal,

$$\{x_n\} = \cup \{x_n : d(x_n,y_k) < R\}.$$

Also, by (5.6),

$$(1 - |x_n|)^{N-1} \leq \text{const.}(1 - |y_k|)^{N-1}$$

when $d(x_n,y_k) < R$. Therefore (5.10) yields

$$\sum_n (1 - |x_n|)^{N-1} \leq \text{Const.} \sum \sigma(E_k) < \infty. \qquad \square$$

Theorem 5 is due to Sullivan [41]; the proof given here is Jones' modification of Thurston's proof in Ahlfors' notes [4]. See [41] and [42] for much fuller discussions of this topic. In [19] the above proof gives the same results for groups of quasiconformal self maps of B^N, $N \geq 3$.

The limit set Σ of a discontinuous group G is the set of all limit points of $\{T_n(0)\}$. Ahlfors' fundamental paper [3] posed the still unresolved problem of showing that $\sigma(\Sigma) = 0$ when G is finitely generated and $N \geq 3$.

6. MAKAROV'S THEOREMS

Let $h(t)$, $t \geq 0$, be continuous and increasing, with $h(0) = 0$. Write $\Delta(z_0, r)$ for the plane disc $\{z : |z - z_0| < r\}$. The Hausdorff measure, with respect to h, of a Borel set E is

$$\Lambda_h(E) = \lim_{\varepsilon \to 0}\left[\inf\left\{\sum h(r_j) : E \subset \bigcup \Delta(z_j, r_j),\ r_j \leq \varepsilon\right\}\right].$$

If $h(t) = t^{\alpha}$, $\alpha > 0$, Λ_h is abbreviated as Λ_{α}. Then $\pi\Lambda_2(E) = \text{Area}(E)$ and, if E lies on a rectifiable curve, $2\Lambda_1(E) = \text{length}(E)$. In 1984 N.G. Makarov obtained some beautiful precise Hausdorff measure estimates of harmonic measures on simply connected domains.

Theorem 6. Let Ω *be simply connected, let* $w_0 \in \Omega$ *and set* $\omega(E) = \omega(w_0, E, \Omega)$, $E \subset \partial\Omega$. *If*

$$\lim_{t \to 0} h(t)/t = 0,$$

then ω *is singular to* Λ_h, $\omega \perp \Lambda_h$.

If $\partial\Omega$ is a Jordan curve Ω is called a *Jordan domain*.

Theorem 7. There are constants C_1 *and* C_2 *such that*

a) *If* $h_1(t) = t \exp\{C_1\sqrt{(\log \frac{1}{t}) \log\log\log \frac{1}{t}}\}$ *then for any Jordan domain* Ω, *harmonic measure on* $\partial\Omega$ *is absolutely continuous to* Λ_{h_1}, $\omega \ll \Lambda_{h_1}$.

b) *If* $h_2(t) = t \exp\{C_2\sqrt{(\log \frac{1}{t}) \log\log\log \frac{1}{t}}\}$ *then there exists a Jordan domain* Ω *such that harmonic measure on* $\partial\Omega$ *is singular to* Λ_{h_2}.

The classical F. and M. Riesz theorem [39] states that if $\partial\Omega$ is a rectifiable Jordan curve, then $\omega \ll \Lambda_1 \ll \omega$. Lavrentiev [25] gave the first example of a Jordan domain for which $\omega \not\ll \Lambda_1$. McMillan and Piranian [31] found a simpler example with $\omega \perp \Lambda_1$, and Kaufman and Wu showed there is a Jordan domain with $\omega \not\ll \Lambda_h$ when $h(t) = t \exp\{A\sqrt{\log \frac{1}{t}}\}$ for any $A > 0$ [24]. For simply connected domains Beurling's projection theorem (see below) yields $\omega \ll \Lambda_{1/2}$, and this was improved by Carleson [12] to $\omega \ll \Lambda_\alpha$ for some constant $\alpha > 1/2$. For general plane domains Øksendal [35] has proved $\omega \perp \Lambda_2$.

Lemma 6.1. *Let D be the unit disc. There is a constant c_0 such that if E is a continuum, if $E \subset \{1/2 \le |z| \le 1\}$, and if E meets both circles $\{|z| = 1/2\}$ and $\{|z| = 1\}$, then*

$$\omega(0, E, D\backslash E) \ge c_0.$$

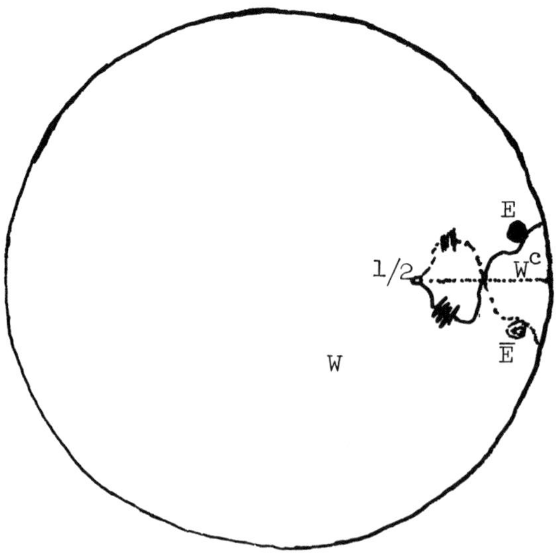

Figure 5: The proof of Lemma 6.1.

Proof. We can assume $1/2 \in E$. Let $\overline{E} = \{\overline{z} : z \in E\}$. Because E is a continuum, the component W of $D\setminus(E \cup \overline{E})$ such that $0 \in W$ satisfies $\partial W \subset E \cup \overline{E} \cup \partial D$ and

$$[1/2,1] \cap W = \emptyset.$$

Set

$$u(z) = \omega(z,E,D\setminus E)$$

$$v(z) = \omega(z,[1/2,1],D\setminus[1/2])$$

Then by the maximum principle,

$$u(z) + u(\overline{z}) \geq v(z), \qquad z \in W.$$

Hence

$$\omega(0,E,D\setminus E) = u(0) \geq \frac{v(0)}{2} = c_0. \qquad \square$$

Proof of Theorem 6. Let φ be a conformal mapping from D onto Ω, $\varphi(0) = w_0$. For $\zeta \in \partial D$, let

$$\Gamma(\zeta) = \{z \in D : |z - \zeta| \leq 2(1 - |z|),\ |z| > 1/2\}$$

be the nontangential cone at ζ. Then

$$\underline{\lim}_{\Gamma(\zeta)\ni z\to\zeta} |\varphi'(z)| < \infty \tag{6.1}$$

at almost every $\zeta \in \partial D$, because on the set where (6.1) fails, the non-constant analytic function $f(z) = 1/\varphi'(z)$ has nontangential limit zero, and by Privalov's theorem such a set has measure zero. Set

$$E_N = \{\zeta \in \partial D : \underline{\lim}_{\Gamma(\zeta)\ni z\to\zeta} |\varphi'(z)| < N\}$$

Then $E_N \subset E_{N+1}$ and by (6.1), $|\cup E_N| = 2\pi$, where $|E|$ denotes the Lebesgue measure of $E \subset \partial D$.

Fix $\varepsilon > 0$. Choose δ_N such that $\delta_N < \varepsilon/N$ and such that if $t < 4N\delta_N$ then

$$h(t) < \frac{\varepsilon t}{4N2^N} \tag{6.2}$$

For $z \in D$, $|z| > 1/2$, let $I(z)$ be the arc $\{\zeta \in \partial D : |\zeta - z| < 2(1 - |z|)\}$, so that $\zeta \in I(z)$ if and only if $z \in \Gamma(\zeta)$. Each $\zeta \in E_N$ lies on arbitrarily small arcs $I(z)$ such that

$$|\varphi'(z)| < N \tag{6.3}$$

and

$$1 - |z| < \delta_N. \tag{6.4}$$

By Vitali's Lemma there are $\{z_{n,N}\}$ satisfying (6.3) and (6.4) such that

$$E_N \subset \cup\, I(z_{n,N}) \tag{6.5}$$

and

$$\sum_n |I(z_{n,N})| \leq C.$$

Then (6.2), (6.3) and (6.4) give us

$$\sum_n h(2|\varphi'(z_{n,N})|(1 - |z_{n,N}|^2)) < C\varepsilon/2^N. \tag{6.6}$$

Because $\delta_N \to 0$, (6.5) implies that $\{z_n\} = \bigcup_N \{z_{n,N}\}$ is <u>nontangentially</u> <u>dense</u>: almost every $\zeta \in \partial D$ is the nontangential limit of a subsequence of $\{z_n\}$.

Now let $w_n = \varphi(z_n)$, $r_n = \operatorname{dist}(w_n, \partial\Omega)$ and $\Delta_n = \Delta(w_n, 2r_n)$. The points w_n depend on $\varepsilon > 0$. We set $A_\varepsilon = \partial\Omega \cap \bigcup_n \Delta_n$ and $A = \bigcap_j A_{1/j}$. We will show that $\Lambda_h(A) = 0$ but that $\omega(w_0, A, \Omega) = 1$.

First consider $\Lambda_h(A)$. By (3.1) from section 3, and by (6.3), (6.4), and the choice of δ_N,

$$2r_n \leq 2|\varphi'(z_n)|(1 - |z_n|^2) < 4\varepsilon,$$

and by (6.6), $\sum_n h(2r_n) < C\varepsilon$. Therefore $\Lambda_h(A) = 0$.

To prove $\omega(w_0, A, \Omega) = 1$ we fix ε and show $\omega(w_0, A_\varepsilon, \Omega) = 1$. Suppose not. Then by the definition (1.1) of harmonic measure

$$u(w) = \omega(w, \partial\Omega \setminus A_\varepsilon, \Omega)$$

satisfies $\sup_{w\in\Omega} u(w) = 1$. Let W_n be that component of $\Delta_n \cap \Omega$ with $w_n \in W_n$. By Lemma 6.1, which applies because $\partial\Omega$ is connected,

$$\begin{aligned} u(w_n) &\le \omega(w_n, \partial\Delta_n \cap \partial W_n, W_n) \\ &\le 1 - c_0. \end{aligned}$$

Hence $v(z) = u(\varphi(z))$ satisfies $v(z_n) \le 1 - c_0$ and $\sup_{z\in D} v(z) = 1$. But that is a contradiction because $\{z_n\}$ is nontangentially dense and because $v(z)$ is the Poisson integral of its nontangential limit. □

We refer to [27] for the proof of Theorem 7, but we shall briefly explain where the triple logarithm comes from. As the proof of Theorem 6 indicates, estimates for harmonic measure in a simply connected domain Ω are equivalent to growth conditions on $\varphi'(z)$, where φ is a conformal mapping from D onto Ω. Suppose $\varphi'(0) = 1$. Then $b(z) = \log \varphi'(z)$ is a Bloch function:

$$\|b\|_B = \sup_{z\in D} (1 - |z|^2)|b'(z)| < \infty, \qquad b(0) = 0.$$

Conversely, if $\|b\|_B$ is small, then $b = \log \varphi'$ where φ maps D conformally onto a Jordan domain. The lacunary series

$$b_0(z) = \sum_{k=0}^{\infty} z^{2^k}$$

is a Bloch function. It resembles a sum of independent identically distributed random variables and for b_0 Salem and Zygmund [40] proved the law of the iterated logarithm:

$$\overline{\lim_{n\to\infty}} \left| \sum_{k=0}^{n} z^{2^k} \right| \Big/ \sqrt{n \log\log n} = 1 \tag{6.7}$$

almost everywhere on $\{|z| = 1\}$. With $r = 1 - e^{-n}$, (6.7) is equivalent to

$$\overline{\lim_{r\to 1}} \frac{|b_0(rz)|}{\sqrt{\log \frac{1}{1-r} \log\log\log \frac{1}{1-r}}} = 1$$

almost everywhere on $\{|z| = 1\}$. A key step in Theorem 7 is Makarov's estimate

$$\overline{\lim_{r\to 1}} \frac{|b(rz)|}{\sqrt{\log \frac{1}{1-r} \log \log \log \frac{1}{1-r}}} \leq C_3 \|b\|_B$$

almost everywhere, for every Bloch function. Determining the best value of C_3 is an open problem.

Beurling's projection theorem is that if E is a compact subset of $D\backslash\{0\}$ and if $E^* = \{|z| : z \in E\}$ is its circular projection onto $(0,1]$, then

$$\omega(0,E,D\backslash E) \geq \omega(0,E^*,D\backslash E^*). \tag{6.8}$$

This theorem clearly implies Lemma 6.1, in which it shows $c_0 = \omega(0,[1/2,1],D\backslash[1/2,1])$ is the best constant. That $\omega \ll \Lambda_{1/2}$ is also immediate from (6.8). Suppose $\infty \in \Omega$ and suppose Ω has diameter 1. For $z_0 \in \partial\Omega$ and for r small, (6.8) applied to $\{|z - z_0| > r\}$ gives

$$\omega(\infty,\partial\Omega \cap \{|z - z_0| \leq r\},\Omega) \leq cr^{1/2},$$

which yields $\omega \ll \Lambda_{1/2}$. See [6] or [2] for the proof of (6.8).

Very recently Jones and Wolff proved that $\omega \perp \Lambda_\alpha$ for all plane domains and for all $\alpha > 1$. Their methods make less use of complex variables and also yield Carleson's theorem [13] that for planar Cantor sets of large Hausdorff dimension there is $\alpha < 1$ such that $\omega \ll \Lambda_\alpha$. For domains in R^N, $N \geq 3$, the relations between harmonic measure and Hausdorff measures are not yet understood.

7. PERIODIC SPECTRA OF HILL'S EQUATION

Let $q(x)$ be a real-valued function in $L^2([0,1])$, extended periodically to $\mathbb{R}$ by $q(x+1) = q(x)$. <u>Hill's equation</u> with potential $q(x)$ is

$$-y'' + q(x)y = \lambda y, \qquad -\infty < x < \infty, \tag{7.1}$$

where λ is any complex constant. By definition, a solution of (7.1) is a function $y(x)$ such that y and y' are absolutely continuous and such that (7.1) holds almost everywhere. Two solutions are given by the integral equations

$$y_1(x,\lambda,q) = \cos\sqrt{\lambda}x + \int_0^x \frac{\sin\sqrt{\lambda}\,(x-t)}{\sqrt{\lambda}}\, y_1(t,\lambda,q)q(t)dt \tag{7.2}$$

and

$$y_2(x,\lambda,q) = \frac{\sin\sqrt{\lambda}x}{\sqrt{\lambda}} + \int_0^x \frac{\sin\sqrt{\lambda}\,(x-t)}{\sqrt{\lambda}}\, y_2(t,\lambda,q)q(t)dt. \tag{7.3}$$

That is, iterating (7.2) and (7.3) yields solutions y_1 and y_2 of (7.1) such that y_1 and y_2 are entire functions of λ satisfying

$$|y_j(x,\lambda,q)| \le \exp(c_1|\mathrm{Im}\sqrt{\lambda}|x)$$

where c_1 depends only on $\|q\|_2 = (\int_0^1 q(t)dt)^{1/2}$. Moreover,

$$\begin{pmatrix} y_1(0) & y_2(0) \\ y_1'(0) & y_2'(0) \end{pmatrix} = \begin{pmatrix} 1 & 0 \\ 0 & 1 \end{pmatrix} \tag{7.4}$$

and every solution of (7.1) has the form

$$y(x) = y(0)y_1(x) + y'(0)y_2(x). \tag{7.5}$$

See [26] or [38] for the proofs of these facts and for more complete discussions of the background outlined below.

The so-called <u>band spectrum</u> of (7.1) is

$$\Sigma = \{\lambda : (7.1) \text{ has a non-zero solution } y(x) \text{ such that } y \text{ and } y' \text{ are bounded on } \mathbb{R}\}.$$

It turns out that Σ is the spectrum of the unique selfadjoint extension to $L^2(\mathbb{R})$ of the operator $-d^2/dx^2 + q(x)$, acting on $C_0^2(\mathbb{R})$. From the general theory of selfadjoint operators it follows that $\Sigma \subset \mathbb{R}$, which we shall soon prove directly.

Floquet theory reduces the description of Σ to a problem in complex analysis. The matrix

$$\mathfrak{F} = \mathfrak{F}(\lambda,q) = \begin{pmatrix} y_1(1,\lambda,q) & y_2(1,\lambda,q) \\ y_1'(1,\lambda,q) & y_2'(1,\lambda,q) \end{pmatrix}$$

is called the Floquet matrix of (7.1). By (7.5) every solution of (7.1) satisfies

$$\begin{pmatrix} y(1) \\ y'(1) \end{pmatrix} = \mathfrak{F} \begin{pmatrix} y(0) \\ y'(0) \end{pmatrix}. \tag{7.6}$$

By (7.4) and the theorem on the Wronskian, $\det \mathfrak{F}(\lambda,q) = 1$ for all λ, so that $\mathfrak{F}$ is invertible. Since $q(x+1) = q(x)$, (7.6) shows that every solution satisfies

$$\begin{pmatrix} y(n+x) \\ y'(n+x) \end{pmatrix} = \begin{pmatrix} y_1(x) & y_2(x) \\ y_1'(x) & y_2'(x) \end{pmatrix} \mathfrak{F}^n \begin{pmatrix} y(0) \\ y'(0) \end{pmatrix} \tag{7.7}$$

for $0 \le x \le 1$ and for all integers n. Consequently, $\lambda \in \Sigma$ if and only if there is a non-zero vector $v = \begin{pmatrix} y(0) \\ y'(0) \end{pmatrix}$ such that the two tailed sequence $\{\mathfrak{F}^n v : -\infty < n < \infty\}$ remains bounded. Write the characteristic equation of $\mathfrak{F}(\lambda,q)$ as

$$\rho^2 - \Delta(\lambda,q)\rho + 1 = 0 \tag{7.8}$$

in which the trace

$$\Delta(\lambda,q) = y_1(1,\lambda,q) + y_2'(1,\lambda,q)$$

is called the discriminant of (7.1). Then (7.8) has roots ρ and $1/\rho$ with $|\rho| \le 1$ and $\rho + 1/\rho = \Delta(\lambda,q)$. Three cases now arise.

Case 1. $\Delta(\lambda,q)$ is not in the real interval $\{-2 \le \Delta \le 2\}$. Then $|\rho| < 1$ and there are independent complex vectors v_1 and v_2 such that for any $n \in \mathbb{Z}$,

$$\left.\begin{aligned} \mathfrak{F}^n v_1 &= \rho^n v_1 \\ \mathfrak{F}^n v_2 &= \rho^{-n} v_2. \end{aligned}\right\} \tag{7.9}$$

Consequently $\mathfrak{F}^n v$ is unbounded for every nonzero v, and $\lambda \notin \Sigma$.

Case 2. $-2 < \Delta(\lambda,q) < 2$. Then $|\rho| = 1$ and $\rho \ne 1/\rho$. There are independent v_1 and v_2 satisfying (7.9) with $|\rho| = 1$, and there are two independent solutions with y and y' bounded. So $\lambda \in \Sigma$, and, in fact, every solution has y and y' bounded.

Case 3. $\Delta(\lambda,q) = \pm 2$. Then $\rho = \pm 1$ and there is a solution either periodic or aperiodic,

$$\begin{pmatrix} y(x+1) \\ y'(x+1) \end{pmatrix} = \pm \begin{pmatrix} y(x) \\ y'(x) \end{pmatrix},$$

and $\lambda \in \Sigma$. The set $\{\Delta(\lambda) = \pm 2\}$ is called the periodic spectrum of (7.1). In passing we note that in this case all solutions are bounded if and only if λ is a double root of $\Delta^2 - 4 = 0$. See [26].

Now if $-2 \leq \Delta(\lambda,q) \leq 2$, then the solution $y(x)$ corresponding to the eigenvalue v_1 satisfies

$$|y(n + x)|^2 = |y(x)|^2, \qquad 0 \leq x \leq 1, \tag{7.10}$$

by (7.7). If $\text{Im}\,\lambda \neq 0$, then

$$[y,\bar{y}] = y\bar{y}' - y'\bar{y}$$

is bounded and, because q is real,

$$[y,\bar{y}]' = (\lambda - \bar{\lambda})|y(x)|^2,$$

contradicting (7.10). Therefore

$$\Sigma \subset \mathbb{R},$$

and we conclude that

$$\Sigma = \{\lambda : -2 \leq \Delta(\lambda,q) \leq 2\} \subset \mathbb{R} \tag{7.11}$$

By (7.2) and (7.3), $\Delta(\lambda,q)$ has the properties

a) $\Delta(\lambda,q)$ is entire in λ,

b) $\Delta(\lambda,q)$ is real for real λ,

c) $\Delta(\lambda,q) = 2\cos\sqrt{\lambda} + (\int_0^1 q(t)dt)\,\dfrac{\sin\sqrt{\lambda}}{\sqrt{\lambda}} + O\left(\dfrac{e^{|\text{Im}\sqrt{\lambda}|}}{|\lambda|}\right)$

d) $\Delta(\lambda,q) \to +\infty \quad (\lambda \to -\infty)$.

Hence by Hadamard's theorem, $\Delta^2(\lambda,q) - 4$ has infinitely many zeros $\{\lambda_n;\ n \geq 0\}$ and by (d) and (7.11), the zeros are a real sequence tending to $+\infty$,

$$\lambda_0 \leq \lambda_1 \leq \lambda_2 \leq \to \infty.$$

By Laguerre's theorem [8], $\dot{\Delta}(\lambda) = \partial\Delta/\partial\lambda$ has one zero between every consecutive pair of zeros of $\Delta(\lambda) - 2$ and $\dot{\Delta}(\lambda)$ has no other zeros. The zeros of $\dot{\Delta}(\lambda)$ also interlace the zeros of $\Delta(\lambda) + 2$ in the same way. This means the values $\Delta(\lambda_n)$ follow the table

λ_n	$\lambda_0 <$	$\lambda_1 \leq$	$\lambda_2 <$	$\lambda_3 \leq$	$\lambda_4 <$	$\lambda_5 \leq$	$\lambda_6 < \cdots$
$\Delta(\lambda_n)$	2	-2	-2	2	2	-2	-2 $\cdots$

and $\Delta(\lambda)$ looks like

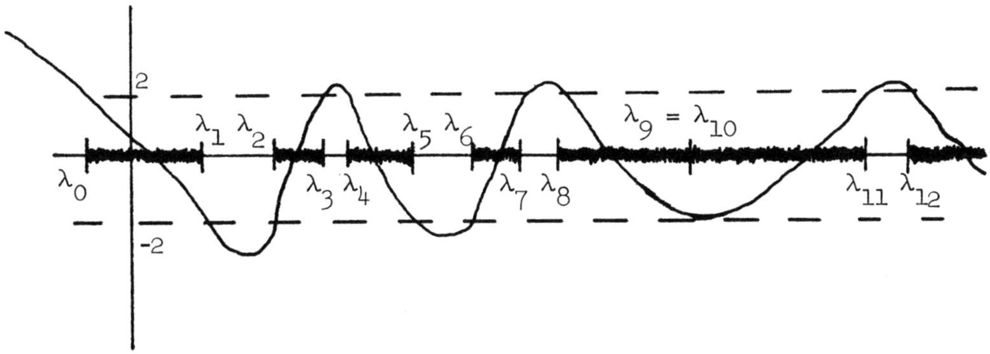

The band spectrum is

$$\Sigma = \bigcup_{n=1}^{\infty} [\lambda_{2n-2}, \lambda_{2n-1}];$$

it is shaded in the figure. The periodic spectrum is the sequence $\{\lambda_n : n \geq 0\}$. The n-th <u>band</u> is the interval $[\lambda_{2n-2}, \lambda_{2n-1}]$ and the n-th <u>gap</u> is the possibly void n-th intervening interval $(\lambda_{2n-1}, \lambda_{2n})$. We want to determine which arrangements of bands and gaps can be spectra. Write

$$\alpha_n = \alpha_n(q) = \lambda_{2n-1} - \lambda_{2n-2}, \qquad n \geq 1$$

for the n-th band length and

$$\gamma_n = \gamma_n(q) = \lambda_{2n} - \lambda_{2n-1}, \qquad n \geq 1$$

for the n-th gap length. Rouché's theorem, and a refinement of the error term in (c), yield the asymptotics

$$\left.\begin{matrix}\lambda_{2n} \\ \lambda_{2n-1}\end{matrix}\right\} = n^2\pi^2 + \int_0^1 q(t)dt + \ell^2$$

where $+\ell^2$ means that the errors are square summable. Consequently

$$\alpha_n = (2n - 1)\pi^2 + \ell^2 \tag{7.13}$$

and

$$\{\gamma_n\} \in \ell^2_+, \tag{7.14}$$

i.e. $\Sigma\,\gamma_n^2 < \infty$ and $\gamma_n \geq 0$.

<u>Example</u>. If $q = 0$, then $y_1 = \cos\sqrt{\lambda}x$, $y_2 = \dfrac{\sin\sqrt{\lambda}x}{\sqrt{\lambda}}$ and $\Delta(\lambda,q) = 2\cos\sqrt{\lambda}$. Hence $\lambda_{2n-1} = \lambda_{2n} = n^2\pi^2$, $\Sigma = [0,\infty)$, $\alpha_n = (2n - 1)\pi^2$, and $\gamma_n = 0$.

Replacing $q(x)$ by $q(x) + c$ translates the spectrum c units because by (7.4) $y_j(x,\lambda + c, q + c) = y_j(x,\lambda,q)$. Thus we also have $\alpha_n = (2n - 1)\pi^2$ and $\gamma_n = 0$ if q is constant. There is a converse due to Borg [9]: if $\gamma_n = 0$ for all n then q is constant.

<u>Theorem 8</u>. <u>For all</u> n,

$$\alpha_n \leq (2n - 1)\pi^2$$

<u>and</u> q <u>is constant if equality holds for any</u> n.

<u>Theorem 9. Whenever</u> $\gamma_n \in \ell^2_+$, <u>that is,</u> $\gamma_n \geq 0$ <u>and</u> $\Sigma \gamma_n^2 < \infty$, <u>there exists a potential</u> $q(x) \in L^2_{\mathbb{R}}([0,1])$ <u>such that</u> $\gamma_n = \gamma_n(q)$ <u>for all</u> n. <u>Moreover, any two band spectra with gap lengths</u> γ_n <u>are translates of one another.</u>

In other words, if a configuration of bands and gaps is a spectrum, then the gap lengths determine the band lengths, or, which is the same, the gap lengths determine the spectrum up to a translation. And any sequence permitted by (7.14) occurs as a sequence of band lengths. We shall not give a description of the isospectral set $\{q \in L^2_{\mathbb{R}}([0,1]) : \gamma_n(q) = \gamma_n\}$, which can be very large. See [30] or [43].

Theorem 8 and Theorem 9 will be proved with harmonic measures. Set

$$\delta(\lambda,q) = \cos^{-1}\left(\frac{\Delta(\lambda,q)}{2}\right) = -\int_{\lambda_0}^{\lambda} \frac{\dot{\Delta}(\zeta)d\zeta}{\sqrt{4 - \Delta^2(\zeta)}}$$

From the above discussion of $\Delta(\lambda,q)$ we see that $\delta(\lambda,q)$ is a conformal mapping from the upper half plane $\mathcal{U} = \{\text{Im}\,\lambda > 0\}$ onto a slit quarter plane

$$\Omega(h) = \{x > 0,\ y > 0\} \setminus \bigcup_{n=1}^{\infty} T_n$$

where

$$T_n = \{n\pi + iy : 0 < y \leq h_n\}$$

and $h_n \geq 0$. The symbol h, standing for the sequence $\{h_n : n \geq 0\}$ of slit heights, parametrizes the set of such domains. Under $\delta(\lambda,q)$ the n-th band $[\lambda_{2n-2},\lambda_{2n-1}]$ is mapped one-to-one onto the interval $\{(n-1)\pi \leq x \leq n\pi,\ y = 0\} \subset \partial\Omega(h)$ and the n-th gap $(\lambda_{2n-1},\lambda_{2n})$ is mapped two-to-one to the n-th slit T_n.

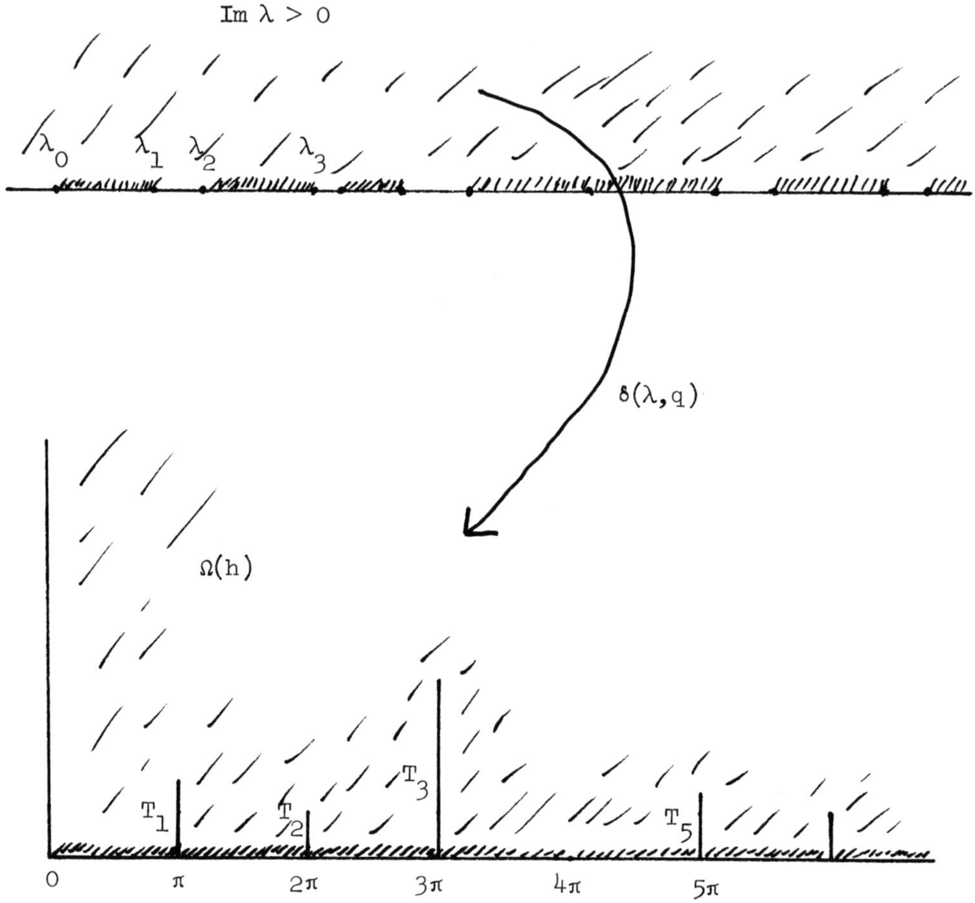

Figure 7: Under $\delta(\lambda,q)$ gaps go to slits. The fourth gap and fourth slit are empty. The map turns left at each simple λ_j, and only there.

The product expansions

$$\dot{\Delta}(\lambda) = \prod_{n=1}^{\infty} \frac{\dot{\lambda}_n - \lambda}{n^2\pi^2},$$

where $\dot{\lambda}_n$ is the zero of $\dot{\Delta}(\lambda)$ in the n-th closed gap $[\lambda_{2n-1},\lambda_{2n}]$, and

$$4 - \Delta^2(\lambda) = 4(\lambda - \lambda_0) \prod_{n=1}^{\infty} \frac{(\lambda_{2n} - \lambda)(\lambda_{2n-1} - \lambda)}{n^4\pi^4},$$

follow from property (c) of $\Delta(\lambda)$. See [38]. With (7.12) they yield the estimate

$$\left| \frac{\dot{\Delta}(\lambda)}{\sqrt{4 - \Delta^2(\lambda)}} \right| \leq \frac{C}{n} \frac{|\dot{\lambda}_n - \lambda|}{\sqrt{(\lambda - \lambda_{2n-1})(\lambda_{2n} - \lambda)}}$$

if $\lambda_{2n-1} \leq \lambda \leq \lambda_{2n}$. Therefore we have

$$h_n \leq \int_{\lambda_{2n-1}}^{\lambda_{2n}} \left| \frac{\dot{\Delta}(\lambda)}{\sqrt{4 - \Delta^2(\lambda)}} \right| d\lambda$$

$$\leq \frac{C}{n} \gamma_n \int_{\lambda_{2n-1}}^{\lambda_{2n}} \frac{d\lambda}{\sqrt{(\lambda - \lambda_{2n-1})(\lambda_{2n} - \lambda)}} \leq \frac{C' \gamma_n}{n}$$

and

$$\Sigma\, n^2 h_n^2 < \infty. \tag{7.15}$$

Because the operation $q \to q - \int_0^1 q(t)dt$ changes the spectrum only by a translation, we can assume $q(x)$ is normalized by

$$\int_0^1 q(t)dt = 0.$$

It then follows from (c) that in $\{\varepsilon < \arg \lambda < \pi - \varepsilon\}$

$$\delta(\lambda,q) = \sqrt{\lambda} + O\left(\frac{1}{\sqrt{\lambda}}\right), \qquad (\lambda \to \infty).$$

When (7.15) holds for $h = \{h_n\}$ there is a unique conformal mapping $\delta(\lambda,h)$ from $\{\text{Im}\,\lambda > 0\}$ onto $\Omega(h)$ such that

$$\delta(\lambda,h) = \sqrt{\lambda} + o\left(\frac{1}{\sqrt{\lambda}}\right), \quad (\lambda \to \infty), \quad \varepsilon < \arg\lambda < \pi - \varepsilon. \tag{7.16}$$

Therefore $\delta(\pi,q)$ is now determined by $\Omega(h)$. Moreover, a theorem of Marčenko and Ostrovskiĭ, [29] or [20], asserts that <u>whenever</u> h <u>satisfies</u> (7.15) <u>there exists a real potential</u> $q \in L^2([0,1])$ <u>with</u> $\int_0^1 q(x)dx = 0$ <u>such that</u> $\delta(\lambda,h) = \delta(\lambda,q)$.

From now on we assume (7.15) and (7.16). Recall that any bounded harmonic function on any $\Omega(h)$ has a nontangential limit almost everywere with respect to arc length on $\partial\Omega(h)$, provided each slit T_n is regarded as two-sided. The next lemma is our tool for estimating lengths of gaps and bands.

<u>Lemma 7.1.</u> <u>Let</u> $u(z)$ <u>be a bounded harmonic function on</u> $\Omega(h)$ <u>such that the nontangential limit of</u> $u(z)$ <u>on</u> $\partial\Omega(h)$ <u>has compact support and let</u>

$$U(\lambda) = u(\delta(\lambda,h)), \qquad \text{Im}\,\lambda > 0.$$

<u>Then the nontangential limit</u> $U(t)$, $t \in \mathbb{R}$, <u>is integrable and</u>

$$\int_{-\infty}^{\infty} U(t)dt = \lim_{x\to+\infty} 2\pi x^2 u(x + ix) \tag{7.17}$$

<u>And if</u> $u(z) > 0$ <u>for all</u> $z \in \Omega(h)$, <u>then</u>

$$\int_{-\infty}^{\infty} U(t)dt > 0.$$

<u>Proof.</u> The boundary function $U(t)$ has compact support because $u(z)$ has compact support on $\partial\Omega(h)$. Hence $\int |U(t)|dt < \infty$ because $U(t)$ is bounded. Then for $\lambda = \xi + i\eta$

$$U(\xi + i\eta) = \frac{\eta}{\pi}\int_{-\infty}^{\infty} \frac{U(t)}{(t-\xi)^2 + \eta^2}\,dt,$$

and dominated convergence gives

$$\int_{-\infty}^{\infty} U(t)dt = \lim_{\eta\to+\infty} \pi\eta U(\xi + i\eta)$$

uniformly in $|\xi| \leq C$. When $\delta(\xi + i\eta, h) = x + ix$ we have by (7.16),

$$\eta = 2x^2 + O(1) \qquad (x \to \infty)$$

$$\xi = O(1) \qquad (x \to \infty),$$

so that

$$\pi\eta U(\xi + i\eta) = 2\pi x^2 u(x + ix) + o(1)$$

That proves (7.17). If $u(z) > 0$ for all $z \in \Omega(h)$, then $U(t) \geq 0$ but $U(\xi + i\eta) > 0$, and we see from the Poisson representation that $\int U(t)dt > 0$. □

<u>Proof of Theorem 8.</u> For any h with (7.15) define

$$u_n(h,z) = \omega(z, B_n, \Omega(h))$$

where $B_n = \{(n - 1)\pi \leq x \leq n\pi,\ y = 0\} \subset \partial\Omega(h)$. When $h = h(q)$, i.e. when $\delta(\lambda,h) = \delta(\lambda,q)$, the lemma gives

$$\alpha_n = \lim_{x\to+\infty} 2\pi x^2 u_n(h, x + ix),$$

and when $h \equiv 0$, i.e. $h_n = 0$ for all n, we have

$$(2n - 1)\pi^2 = \lim_{x\to\infty} 2\pi x^2 u_n(0, x + ix).$$

If q is not constant then by Borg's theorem $\gamma_n > 0$ for some n. Then the slit T_n is not empty and $h_n > 0$. On $\Omega(h)$ the difference $u(0,z) - u(h,z)$ satisfies the hypothesis of the lemma and

$$u(0,z) - u(h,z) > 0$$

by the maximum principle. Then the second assertion of the lemma yields

$$(2n - 1)\pi^2 - \alpha_n = \lim 2\pi x^2(u(0,x + ix) - u(h,x + ix)) \geq 0.$$

<u>Proof of Theorem 9</u>. Here we restrict ourselves to the finite case,

$$h_n = 0, \quad n > N$$

and refer to [20] for the approximations leading to the general case. Write

$$\gamma_n(h) = \text{length}(\delta(\lambda,h)^{-1}(T_n))$$
$$\gamma(h) = (\gamma_1(h),\ldots,\gamma_N(h))$$

and

$$\overline{R^N_+} = \{x \in R^N : x_n \geq 0,\ 1 \leq n \leq N\}.$$

By the Marčenko-Ostrovskiĩ theorem, the finite case of Theorem 9 is equivalent to the assertion that

$$\overline{R^N_+} \xrightarrow{\gamma(h)} \overline{R^N_+}$$

is one-to-one and onto.

<u>Lemma 7.2</u>. <u>For all</u> n,

$$\gamma_n \leq 4 \operatorname{Max}(2\pi n h_n, h_n^2) \tag{7.18}$$

<u>and there is</u> $C(M,N)$ <u>such that if</u> $\{\gamma_n\} \in \overline{R^N_+}$ <u>and if</u> $\operatorname{Max}\{\gamma_n : 1 \leq n \leq N\} \leq M$ <u>then</u>

$$\max_{1 \leq n \leq N} h_n \leq C(M,N). \tag{7.19}$$

<u>Proof</u>. Let $\omega_n(h,z) = \omega(z,T_n,\Omega(h))$. Then by Lemma 7.1,

$$\gamma_n(h) = \lim_{x\to\infty} 2\pi x^2 \omega_n(h, x + ix).$$

By a comparison, $\omega_n(h,z)$ is increasing in h_n and decreasing in h_k, $k \neq n$, so that $\gamma_n(h) \leq \gamma_n(h')$ where $h'_k = \delta_{k,n}h_n$. Thus we suppose $h_k = 0$, $k = n$. Then $z \to z^2$ maps $\Omega(h)$ onto $\mathcal{U}\backslash\Gamma_n$, where $\mathcal{U}$ is the upper half plane and Γ_n is the parabolic arc

$$\Gamma_n = \{(n^2\pi^2 - s^2) + 2\pi ins;\ 0 \leq s \leq h_n\},$$

and $\omega_n(z,h) = W_n(\sqrt{z})$ where W_n is the harmonic measure of Γ_n in $\mathcal{U}\backslash\Gamma_n$. Enclose Γ_n in the closed disc D_n with center $n^2\pi^2 - h_n^2$ and radius

$$r_n = \mathrm{Max}(2\pi nh_n, h_n^2),$$

and let W'_n be the harmonic measure of $\mathcal{U} \cap \partial D_n$ in $\mathcal{U}\backslash D_n$. Then

$$W'_n(\xi + i\eta) = \frac{2}{\pi} \int_{\mathbb{R}\cap D_n} \frac{\eta}{(\xi - t)^2 + \eta^2}\, dt$$

and $W_n \leq W'_n$. Hence by Lemma 7.1,

$$\begin{aligned}\gamma_n(h) &\leq \lim_{\eta\to\infty} \pi\eta W_n(i\eta)\\ &\leq \lim_{\eta\to\infty} \pi\eta W'_n(i\eta) = 4r_n,\end{aligned}$$

and (7.18) holds. This proof is due to P. Jones.

To prove (7.19) we first consider h_N. Comparing $\omega_N(h,z)$ to the harmonic measure of $\{N\pi + iy : 0 \leq y \leq h_N\}$ in the quarter plane $\{x > N\pi,\ y > 0\}$ and applying the proof of Lemma 7.1 in that quarter plane, we obtain

$$\gamma_N(h) \geq h_N^2.$$

To estimate h_{N-1}, repeat the argument in the quarter plane $\{x > (N - 1)\pi,\ y > h_N\}$. After N steps we get (7.19). □

The map $\gamma(h)$ is real analytic on $R_+^N = \{x \in R^N : x_n > 0, 1 \leq n \leq N\}$. See page 290 of [20] for a proof.

Lemma 7.3. *The Jacobian* $\partial\gamma/\partial h$ *is invertible. Moreover*

$$\frac{\partial\gamma_n}{\partial h_n} \geq 0, \qquad \frac{\partial\gamma_k}{\partial h_n} \leq 0, \qquad k \neq n \tag{7.20}$$

and

$$\frac{\partial\gamma_n}{\partial h_n} + \sum_{k,k\neq n} \frac{\partial\gamma_k}{\partial h_n} > Cne^{-(h^*+h_n)/2} \tag{7.21}$$

where $h^* = \text{Max}\{h_k : k = 1,2,\ldots\}$.

Proof. The Jacobian is invertible because by (7.20) and (7.21) the diagonal entry of each column dominates the absolute sum of the rest of that column.

Since $\omega_n(h,z) = \omega(z,T_n,\Omega(h))$ increases when h_n increases but decreases when h_k increases, $k \neq n$, (7.20) holds.

We turn to the proof of the key inequality (7.21). Fix n, let e_n be the unit vector $(e_n)_j = \delta_{n,j}$ and let $t > 0$. We must find a lower bound for the positive harmonic function

$$V_t(h,z) = \frac{1}{t}\sum_{k=1}^{N} \{\omega_k(h + te_n,z) - \omega_k(h,z)\}$$

on $\Omega(h + te_n)$. Let I_t denote the segment

$$I_t = \{n\pi + i(h_n + s) : 0 < s \leq t\}$$

and let $\Omega_n^*(b)$, $b > 0$, be the slit strip

$$\Omega_n^*(b) = \{|x - n\pi| < \pi, y > 0\}\backslash\{n\pi + iy : 0 < y \leq b\}$$

with base $B_n = \{(n - 1)\pi < x < (n + 1)\pi\}$. When $\zeta \in \partial\Omega_n^*(h_n + t)$ we have

$$V_t(h,\zeta) \geq \frac{1}{t}\left(1 - \sum_{k=1}^{N} \omega_k(h,\zeta)\right) \chi_{I_t}(\zeta)$$
$$\geq \frac{1}{t}\, \omega(\zeta, B_n, \Omega_n^*(h_n))\chi_{I_t}(\zeta),$$

where χ_E indicates the characteristic function of E. To estimate $\omega(\zeta, B_n, \Omega_n^*(h_n))$ for $\zeta \in I_t$ we consider two cases.

Case 1. $h_n + t \leq \pi/2$.
In terms of the coordinate $\omega = (z/\pi) - n$, the slit half disc

$$D_n = \{|w| < 1,\ \text{Im } w > 0\}\setminus\{iy : 0 < y < h_n/\pi\}$$

with diameter B_n has

$$I_t \subset D_n \subset \Omega_n^*(h_n).$$

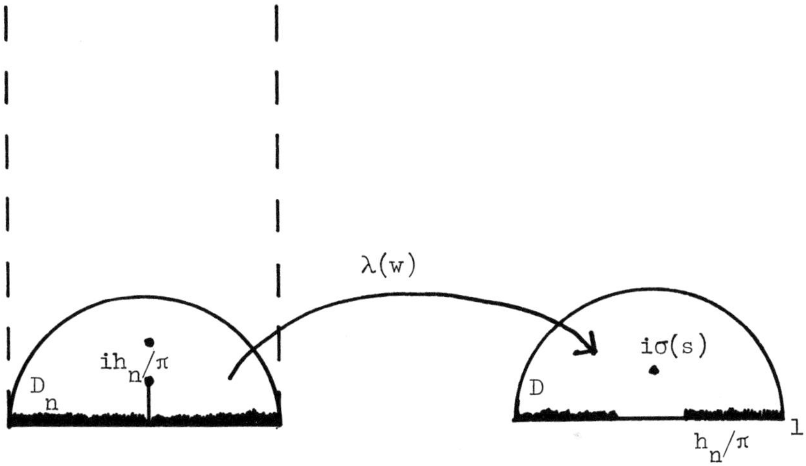

Figure 8

Under

$$\tau(w) = \left\{\frac{\pi^2 w^2 + h_n^2}{\pi^2 + h_n^2 w^2}\right\}^{1/2},$$

D_n is mapped to the half disc $D = \{|\tau| < 1,\ \text{Im}\ \tau > 0\}$ so that

$$\tau(B_n) = [-1, -h_n/\pi] \cup [h_n/\pi, 1]$$

and so that $\zeta(s) = n\pi + i(h_n + s)$ has

$$\tau \circ w(\zeta(s)) = i\pi\left\{\frac{s^2 + 2h_n s}{\pi^4 - h_n^2(h_n + s)^2}\right\} = i\sigma(s)$$

Consequently

$$\omega(\zeta(s), B_n, \Omega_n^*(h_n)) \geq \omega(i\sigma(s), \tau(B_n), D)$$
$$\geq \text{Const.}\left(\frac{s}{h_n}\right)^{1/2} + O\left(\left(\frac{s}{h_n}\right)^{3/2}\right).$$

Case 2. $h_n + t \geq \pi/2$.
On the two horozontal segments $\{\pi/4 < |x - n\pi| < 3\pi/4,\ y = h_n - 1\}$ we have $\omega(z, B_n, \Omega_n^*(h_n)) \geq \text{const.}\ e^{-h_n}$, from a comparison in the strips $\{n\pi < x < (n + 1)\pi,\ y > 0\}$ and $\{(n - 1)\pi < x < n\pi,\ y > 0\}$. Now we can repeat the argument of Case 1 using the slit half disc

$$D_n = \{|z - (n\pi + i(h_n - 1))| < \pi,\ y > h_n - 1\}\backslash T_n$$

to get

$$\omega(\zeta(s), B_n, \Omega_n^*(h_n)) \geq \text{Const.}\ e^{-h_n} s^{1/2}.$$

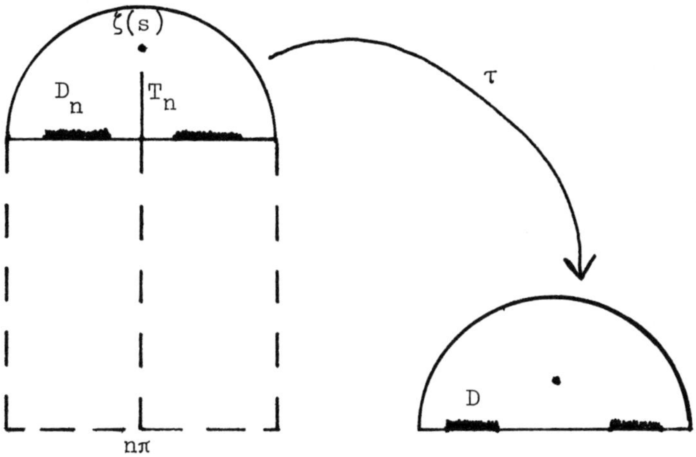

Figure 9

Together the two cases yield

$$\omega(\zeta(s), B_n, \Omega_n^*(h_n)) \geq A(h_n)s^{1/2} \tag{7.23}$$

with

$$A(h_n) = \begin{cases} \text{Const. } e^{-h_n}, & h_n \text{ large} \\ \text{Const.}/h_n^{1/2}, & h_n \text{ small.} \end{cases}$$

Now fix $z_n = n\pi + i(h_n + 1)$, let $0 < t < h_n/2$ and set

$$g(t) = \omega(z_n, I_t \backslash I_{t/2}, \Omega_n^*(h_n + t)).$$

If $h_n \leq \pi/2$, a comparison with the slit half disc $\{|z - n\pi| < \pi\} \backslash \{n\pi + iy : 0 < y < h_n + t\}$ gives

$$g(t) \geq \text{Const.}(h_n t)^{1/2},$$

while if $h_n > \pi/2$, a comparison to the slit disc

$$\{|z - (n\pi + ih_n)| < \pi/2\}\backslash\{n\pi + iy : h_n - \pi/2 < y < h_n + t\}$$

yields $g(t) \geq \text{Const.}(t)^{1/2}$. Therefore in both cases we have, by (7.22) and (7.23),

$$\begin{aligned} V_t(h,z) &\geq \frac{1}{t} \inf_{I_t \backslash I_{t/2}} \omega(\zeta, B_n, \Omega^*(h_n))g(t) \\ &\geq \text{Const. } e^{-h_n}. \end{aligned}$$

Harnack's inequality gives the same lower bound, with a smaller constant, on the disc $\{|z - z_n| < 1/2\}$ and a comparison with the strip

$$\{|x - n\pi| < \pi,\ y > h_n + 1\}$$

then yields

$$V_t(h,z) \geq \text{Const. } e^{-(y+h_n)/2}$$

on $\{|x - n\pi| < \pi/2,\ y > h_n + 1\}$.

Finally let W be the quarter plane $\{x > 0,\ y > 1 + h^* = 1 + \text{Max } h_n\}$. Applying Lemma 7.1 to W, we have

$$\lim_{x\to\infty} 2\pi x^2 V_t(h, x + i(x + h^*)) \geq \text{Const. } ne^{-(h^*+h_n)/2}$$

which implies

$$\lim_{x\to\infty} 2\pi x^2 V_t(h, x + ix) \geq \text{Const. } ne^{-(h^*+h_n)/2}.$$

That proves (7.21) and Lemma 7.3. □

We may now complete the proof of the finite case of Theorem 9. To show the surjectivity of

$$\overline{\mathbb{R}^N_+} \xrightarrow{\gamma(h)} \overline{\mathbb{R}^N_+}$$

first consider the open quadrant $R_+^N = \{x \in R^N : x_n > 0\}$. On R_+^N the map γ is proper by Lemma 7.2 and its Jacobian is invertible by Lemma 7.3. Hence its range is both open and closed in R_+^N, and so γ maps R_+^N onto R_+^N. Now since $h_n = 0$ if and only if $\gamma_n = 0$, the same reasoning applies when $h_k = 0$ on some subset of $\{1,2,\ldots,N\}$.

To prove γ is one-to-one, fix distinct h and $\tilde{h} \in \overline{R_+^N}$ and let α be the unit vector

$$\alpha = \frac{h - \tilde{h}}{\|h - \tilde{h}\|} = (\alpha_1,\ldots,\alpha_N)$$

where $\| \ \|$ is the euclidean norm in R^N. Set

$$\gamma^{(\alpha)} = \sum_{\alpha_n \neq 0} \frac{\alpha_n}{|\alpha_n|} \gamma_n.$$

Then by (7.21),

$$\frac{d}{dt} \nu^{(\alpha)}(h + t(\tilde{h} - h)) > 0$$

on $0 \leq t \leq 1$, and thus $\nu(h) \neq \nu(\tilde{h})$. □

The situation with band lengths is less clear. In particular, I cannot prove the map $\alpha^{(N)} : \overline{R_+^N} \to R^N$, defined by

$$\alpha^{(N)}(h) = (\alpha_1(h),\ldots,\alpha_N(h)),$$

is one-to-one. However, a finite $h \in \overline{R_+^N}$ is determined by the infinite sequence $\{\alpha_n(h) : n \geq 1\}$. The proof is easy.

<u>Theorem 7.4.</u> <u>If</u> $N < \infty$, <u>if</u> $h \in \overline{R_+^N}$ <u>and</u> $\tilde{h} \in \overline{R_+^N}$, <u>and if</u> $\alpha_n(h) = \alpha_n(\tilde{h})$ <u>for all large</u> n, <u>then</u> $h = \tilde{h}$.

<u>Proof</u>. Let $\varphi_h(z)$ be the inverse of $\delta(\lambda,h)$, and set $\psi_h(z) = \varphi_h(z + \pi) - \varphi_h(z)$. If h is finite, then ψ_h is analytic at $z = \infty$, and $\psi_h(n\pi) = \alpha_n(h)$, $n > N$. Thus if $\alpha_n(h) = \alpha_n(\tilde{h})$ for large n then $\psi_h = \psi_{\tilde{h}}$. But if $h \neq \tilde{h}$,

then by considering the largest n for which $h_n \neq \tilde{h}_n$, we find different singularities for ψ_h and $\psi_{\tilde{h}}$. □

The range of the band length map $\alpha_n(h)$ has not been fully described, but the condition

$$\alpha_n \le (2n - 1)\pi^2 \tag{7.24}$$

of Theorem 8 is not the only constraint.

<u>Theorem 7.5.</u> <u>For all</u> $q \in L^2_{\mathbb{R}}([0,1])$ <u>and for all</u> n,

$$\alpha_n - \alpha_{n-1} + \alpha_{n-2} \mp \cdots > 0. \tag{7.25}$$

<u>Moreover, if</u> $\beta_n = (2n - 1)\pi^2 - \alpha_n$, <u>then</u>

$$\tilde{\beta}_n = \beta_n - \beta_{n-1} + \beta_{n-2} \mp \cdots \ge 0. \tag{7.26}$$

Since $\alpha_n = (2n - 1)\pi^2 + \ell^2$, (7.25) has content only for small n. By (7.13) $\beta_n \in \ell^2$ and by (7.24),

$$0 \le \beta_n < (2n - 1)\pi^2.$$

Hence (7.26) implies $\tilde{\beta}_n \in \ell^2$ and (7.25) implies

$$0 \le \tilde{\beta}_n < n\pi^2. \tag{7.27}$$

For fixed n (7.27) is sharp because for $k \le n$, $\alpha_k \to 0$ as $h_n \to \infty$. Of course (7.26) also implies

$$\alpha_n - \alpha_{n-1} + \alpha_{n-2} \mp \cdots \le n\pi^2.$$

It turns out [21] that, suitably defined, the Jacobian $(\partial\tilde{\beta}_n/\partial h_k)$ is invertible, which makes further global linear constraints on $\alpha_n(h)$ unlikely. We conclude with a Brownian motion proof of Theorem 7.5. See [37] for the assumed probability.

Proof. Fix h, and set

$$u_n(z) = \omega(z, B_n, \Omega(h))$$

where $B_n = \{(n-1)\pi < x < n\pi,\ y = 0\} \subset \partial\Omega(h)$. By Lemma 7.1, (7.25) will hold if

$$u_n(z) - u_{n-1}(z) + u_{n-2}(z) \mp \cdots > 0$$

when $z \in \Omega(h) \cap \{\text{Re } z > n\pi\}$. Fix such a point z. Let S_k be the set of Brownian paths beginning at z and first leaving $\Omega(h)$ in B_k. Then by Kakutani's theorem

$$u_k(z) = P_z(S_k).$$

We can assume all paths in S_k are continuous. Set $J_j = \Omega(h) \cap \{y = j\pi\}$. Since $k\pi \leq n\pi < \text{Re } z$, every path in S_k meets J_k before it leave $\Omega(h)$ in B_k. Let L_k denote the set of paths in S_k which, before departing $\Omega(h)$ through B_k, hit J_{k-1} after their last contact with J_k, and let $R_k = S_k \setminus L_k$. Thus a path is in R_k if it first leaves $\Omega(h)$ in a segment joining J_k to B_k without touching J_{k-1}. Now R_k and L_k are measurable and $P_z(S_k) = P_z(R_k) + P_z(L_k)$. But by the symmetry of Brownian motion

$$P_z(L_k) = P_z(R_{k-1}), \qquad 2 \leq k \leq n.$$

Also $L_0 = \emptyset$, since $\{iy : y \geq 0\} \subset \partial\Omega(h)$. Therefore the alternating sum telescopes into $P_z(R_n) > 0$, and (7.25) is proved. The proof of (7.26) is the same, but with a different harmonic function. □

Theorem 8 and Theorem 9 are from [20]. See Moser [32] for a different proof of Theorem 8. See [28], [29], and [20] for two different proofs of the Marčenko-Ostrovskiĭ theorem. A little more about band lengths can be found in [21].

APPENDIX

Let Ω be a plane domain such that every point of $\partial\Omega$ is a regular point for the Dirichlet problem. For $f(\zeta)$ any bounded function on $\partial\Omega$, set

$$\mathcal{V}_f = \{v(z) : v \text{ subharmonic on } \Omega, \lim_{z\to\zeta} v(z) \leq f(\zeta),$$
$$\text{all } \zeta \in \partial\Omega\}$$

and

$$U_f(z) = \sup\{v(z) : v \in \mathcal{V}_f\}.$$

Then U_f is harmonic on Ω and if $f(\zeta)$ is continuous on $\partial\Omega$, then at all $\zeta \in \partial\Omega$,

$$\lim_{z\to\zeta} U_f(z) = f(\zeta)$$

because every ζ is a regular point; see Ahlfors [1, p. 248]. By the Riesz theorem there is a Borel probability measure μ_z on $\partial\Omega$ such that for continuous $f(\zeta)$

$$U_f(z) = \int f(\zeta)d\mu_z(\zeta).$$

We shall show that whenever E is a Borel subset of $\partial\Omega$,

$$\mu_z(E) = \omega(z,E,\Omega) \tag{A.1}$$

where $\omega(z,E,\Omega) = U_{\chi_E}(z)$ is the definition (1.1) of harmonic measure.

Suppose E is a closed subset of $\partial\Omega$. Let $\{f_n(\zeta)\}$ be a sequence of continuous functions decreasing to $\chi_E(\zeta)$. Then

$u_n(z) = U_{f_n}(z)$ is a decreasing sequence of positive harmonic functions having limit a harmonic function $u(z)$. Clearly $u(z) = \mu_z(E)$. Since

$$\overline{\lim_{z\to\zeta}}\, u(z) \le \overline{\lim_{z\to\zeta}}\, u_n(z) = f_n(\zeta)$$

for all n, we have $u \in \mathcal{U}_{\chi_E}$ and hence

$$u(z) \le U_{\chi_E}(z).$$

On the other hand, if $v \in \mathcal{U}_{\chi_E}$, then $v \in \mathcal{U}_{f_n}$ so that $v(z) \le u_n(z)$ for all n. Hence

$$U_{\chi_E}(z) \le \lim_n u_n(z) = u(z),$$

and

$$U_{\chi_E}(z) = u(z) = \mu_z(E)$$

when E is closed.

Since $f \le g$ implies $U_f \le U_g$, we now have

$$\mu_z(E) = \sup\{\mu_z(K) : K \text{ closed, } K \subset E\} \le U_{\chi_E}(z)$$

for all Borel E. And since $U_f + U_g \le U_{f+g}$ (because $\mathcal{U}_f + \mathcal{U}_g \subset \mathcal{U}_{f+g}$), we also have

$$\mu_z(E) = 1 - \mu_z(E') \ge 1 - U_{\chi_{E'}}(z) \ge U_{\chi_E}(z),$$

and (A.1) is proved.

REFERENCES

1. Ahlfors, L. V. Complex Analysis, 2nd edition, McGraw-Hill, New York, 1966.
2. Ahlfors, L. V. Conformal Invariants, Topics in Geometric Function Theory, McGraw-Hill, New York, 1973.
3. Ahlfors, L. V. "Finitely Generated Kleinian Groups," Amer. J. Math. 86 (1964) 413-429.
4. Ahlfors, L. V. Möbius Transformations in Several Dimensions, School of Mathematics, University of Minnesota, 1981.
5. Barth, K. F., Brannan, D. A. and Hayman, W. K. "Research Problems in Complex Analysis" Technical Report No. M-8, Dept. of Mathematics, Syracuse University, Syracuse, N.Y. 1984.
6. Beurling, A. "Etudes sur un problème de majoration" Thèse, Uppsala, 1933.
7. Beurling, A. "Sur les fonctions limites quasianalytique des fonctions rationelles," VIII Congrés des Mathématiciens Scandinaves, Stockholm, 1934.
8. Boas, R. P. Entire Functions, Academic Press, Inc., New York, 1954.
9. Borg, G. Eine Umkehrung der Sturm-Liouvilleschen Eigenwertaufgabe, Acta Math. 78 (1946), 1-96.
10. Brennan, J. "The Integrability of the Derivative in Conformal Mapping," J. London Math. Soc. (2) 18 (1978) 261-272.
11. Carleson, L. "Estimates on Harmonic Measure," Ann. Acad. Sci. Fenn. Ser. A 7 (1982) 25-32.
12. Carleson, L. "On the Distortion of Sets on a Jordan Curve under Conformal Mapping," Duke Math. J. 40 (1973) 547-559.
13. Carleson, L. "On the Support of Harmonic Measure for Sets of Cantor Type," Institut Mittag-Leffler Report No. 4, 1984.
14. Carleson, L. and Garnett, J. "Interpolating Sequences and Separation Properties," J. Analyse Math. 28 (1975) 273-299.
15. Durrett, R. Brownian Motion and Martingales in Analysis, Wadsworth, Belmont, California, 1984.
16. Fuchs, W. H. J. Topics in the Theory of Functions of One Complex Variable, Van Nostrand, Princeton, N.J., 1967.

17. Garnett, J. B. Bounded Analytic Functions, Academic Press, New York, 1981.

18. Garnett, J. B., Gehring, F. W. and Jones, P. W. "Conformally Invariant Length Sums," Indiana U. Math. J. 32 (1983) 809-824.

19. Garnett, J. B., Gehring, F. W. and Jones, P. W. "Quasiconformal Groups and the Conical Limit Set," to appear.

20. Garnett, J. and Trubowitz, E. "Gaps and Bands of One Dimensional Periodic Schrödinger Operators," Comment. Math. Helvetici 59 (1984) 258-312.

21. Garnett, J. and Trubowitz, E. "Gaps and Bands of One Dimensional Periodic Schrödinger Operators, II," to appear.

22. Hall, T. "Sur la mesure harmonique de certains ensembles," Ark. Mat. Astr. Fys. 25A, No. 28 (1937).

23. Hayman, W. and Wu, J. M. G. "Level Sets of Univalent Functions," Comment. Math. Helvetici 56 (1981) 366-403.

24. Kaufman, R. and Wu, J. M. G. "Distortion of the Boundary under Conformal Mapping," Mich. Math. J. 29 (1982) 267-280.

25. Lavrentiev, M. A. "Boundary Problems on the Theory of Univalent Functions," Amer. Math. Soc. Transl. 32 (1963) 1-35.

26. Magnus, W. and Winkler, S. Hill's Equation, Interscience, New York, 1966.

27. Makarov, N. "On the Distortion of Boundary Sets under Conformal Mappings," Proc. London Math. Soc. (3) 51 (1985) 369-384.

28. Marčenko, V. A. Sturm-Liouville Operators and Applications, Kiev 1977 (in Russian).

29. Marčenko, V. A. and Ostrovskiĩ, I. V. "A Characterization of the Spectrum of Hill's Operator," Math. USSR-Sbornik, 97 (1975) 493-554.

30. McKean, M. P. and Trubowitz, E. "Hill's Operator and Hyperelliptic Function Theory in the Presence of Infinitely Many Branch Points," Comm. Pure Appl. Math. 29 (1976) 143-226.

31. McMillan, J. and Piranian, G. "Compression and Expansion of Boundary Sets," Duke Math. J. 40 (1973) 599-605.

32. Moser, J. Integrable Hamiltonian Systems and Spectral Theory, Lezioni Fermiane, Accademia Nazionale dei Lincei

Scuola Normale Superiore, Pisa, 1981.

33. Nevanlinna, R. "Über eine Minimumaufgabe in der Theorie der konformen Abbildung," Nachr. Ges. Wiss. Göttinger, F. I37 (1933) 103-115.

34. Øksendal, B. "Brownian Motion and Sets of Harmonic Measure Zero," Pac. J. Math. 95 (1981) 179-192.

35. Øksendal, B. "Null Sets for Measures Orthogonal to R(X)," Amer. J. Math. 94 (1972) 331-342.

36. Petersen, K. E. Brownian Motion, Hardy Spaces and Bounded Mean Oscillation, Cambridge University Press, Cambridge, 1977.

37. Port, S. and Stone, C. Brownian Motion and Classical Potential Theory, Academic Press, New York, 1978.

38. Pöschel, J. and Trubowitz, E. Lectures on Inverse Spectral Theory, to appear.

39. Riesz, F. and Riesz, M. "Über Randwerte einer analytischen Funktionen," Quartrieme Congrès des Math. Scand. Stockholm (1916) 27-44.

40. Salem, R. and Zygmund, A. "La loi du logarithme itéré pour les séries trigonometric lacunaire," Bull. Sci. Math. 74 (1980) 209-224.

41. Sullivan, D. "On the Ergodic Theory at Infinity of an Arbitrary Discrete Group of Hyperbolic Motions," Riemann Surfaces and Related Topics, Proceedings of the 1978 Stony Brook Conference, Princeton University Press, Princeton, 1978.

42. Thurston, W. P. The Geometry and Topology of Three-Manifolds, Department of Mathematics, Princeton University, Princeton, N.J.

43. Trubowitz, E. "The Inverse Problem for Periodic Potentials," Comm. Pure and Appl. Math. 30 (1977) 321-337.

44. Tsuji, M. Potential Theory in Modern Function Theory, Maruzen, Tokyo, 1959.